*

The Star of
Bethlehem

The Star of
𝕭ethlehem

THE LEGACY OF THE MAGI

MICHAEL R. MOLNAR

RUTGERS UNIVERSITY PRESS
NEW BRUNSWICK, NEW JERSEY, AND LONDON

First paperback printing, 2013
Library of Congress Cataloging-in-Publication Data

Molnar, Michael R., 1945–
  The Star of Bethlehem : the legacy of the Magi / Michael R.
Molnar.
    p.   cm.
  Includes bibliographical references and index.
  ISBN 978-0-8135-6471-5 (pbk : alk. paper)
  ISBN 0-8135-2701-5 (alk. paper)
  1. Star of Bethlehem.   2. Astrology, Greek.   I. Title.
BT315.2.M6   1999
232.92' 3—dc21   98-55376
                                                                    CIP

British Cataloging-in-Publication data for this book is available from the British
Library.

To my wife, Shelley,
my son, Greg,
and my parents

Contents

List of Illustrations ix
List of Tables xi
Preface and Acknowledgments xiii

1 Introduction 1
2 A Mysterious Star 15
3 The Dawning of the First Millennium 32
4 Secrets of Regal Horoscopes 64
5 Astrological Portents for Judea 85
6 Epilogue 124

Appendices 127
  A Defining the Position of the Zodiac 127
  B Astrological Influences 128
  C The System of Mundane Houses and Cardinal Points 132
  D Regal Horoscopes 134
Chronology 139
Notes 145
Glossaries 163
  Astrological and Astronomical Terms 163
  Historical and Numismatic Terms 169
  Significant Historical Persons 170
Bibliography 175
Index 181

Illustrations

FIGURES

1. The Nativity, a woodcut by Albrecht Dürer, from "The Little
   Passion," 1508–1510   2

2. A silver denarius issued by Augustus Caesar, 19–18 B.C.   19

3. Kepler's supernova, 1604   23

4. The Magi shown in eastern attire in the Epitaph of Severa,
   3d–4th century A.D.   37

5. A silver cistophorus bearing Augustus's sign, Capricorn,
   27–26 B.C.   41

6. The zodiac inclined on the celestial sphere   43

7. The shift of the seasons through the zodiacal signs caused by the
   precession of the earth's axis   45

8. One of the first bronze coins of Antioch depicting Aries,
   A.D. 5–11   51

9. The last coin issued by the governor Silanus showing Aries,
   A.D. 13–14   52

10. The benchmarks used to estimate the date of the birth of Jesus   63

11. The Greek manuscript of the emperor Hadrian's horoscope,
    2d century A.D.   66

12. A modern representation of Antigonus's horoscope for Emperor
    Hadrian   67

13. The planets in the locations of their exaltations   68

14. The four trines   70

15. The rulers of Trine I in Aries ♈   71

16. The Sun, the Moon, and Jupiter in the Ascendant in Hadrian's
    horoscope   73

17. A regal portent for Judea   76

18. The Regal Lion Horoscope of Antiochus I at Nemrud Daği   78
19. Retrograde motion and stationing   91
20. The positions of the planets on April 17, 6 B.C.   94
21. A secondary regal portent   95
22. Sunrise for April 17, 6 B.C.   97
23. Jupiter at the Midheaven on April 17, 6 B.C.   98
24. Mars ♂ maligning the regal portent of April 4, A.D. 54   103
25. A bronze coin of Constantine the Great depicting Sol Invictus,
    A.D. 317   108
26. A bronze coin of Quadratus, A.D. 55–56   111
27. A reconstruction of Nero's horoscope for sunrise on
    December 15, A.D. 37   112
28. A bronze coin showing Nero with the star and crescent symbol,
    ca. A.D. 66–68   114
29. The vernal equinox in the fifth degree of Aries   128
30. The domiciles   130
31. A reconstruction of Augustus Caesar's natal horoscope,
    September 23, 63 B.C.   135

MAPS

1. The Kingdom of Herod the Great   4
2. Historical Lands and Cities   9

Tables

1.1.	*Matthew 2:1–16*	12
2.1.	*Cometary Portents Close to the Time of Jesus' Birth*	21
3.1.	*Antioch's First Coins Bearing Aries*	50
3.2	*Roman Governors of Syria*	59
5.1.	*Greek Text of Matthew 2:9*	90
A.1.	*General Astrological Characteristics of the Celestial Bodies*	131
A.2.	*Nature, Sect, and Gender of the Luminaries and Planets*	131
A.3.	*Gender and Sect of the Zodiacal Signs*	132
A.4.	*The System of Mundane Houses*	133

Preface

At Christmas time, television and radio audiences hear the compelling story about how I, a Rutgers University astronomer, came to explain the special star that marks this religious holiday. Like many other astronomers, I would get questions about the biblical report of a "star in the east", the Star of Bethlehem. I said that this was just an unexplainable mystery of the Bible. Moreover, I preferred not being involved with a matter of Christian faith and evaded those questions. However, as I explain "the story behind the story" in those media interviews, the possibility that there was a real historical basis to this biblical account became credible for me over time. One finding followed by another prodded me on to uncover what had truly appeared in the skies that Christians believe to have marked the birth of Jesus.

The story begins with my hobby of collecting ancient coins. Roman and Greek coins can be rather expensive so I narrowed my collecting interest just to coins with celestial symbols such as stars, crescent moons, and signs of the zodiac—a theme befitting an astronomer. Purchasing one particular Roman coin would lead me unexpectedly on an amazing adventure. On one side of the coin appeared Aries the Ram, an astrological sign of the zodiac. I paid $50 for the bronze coin and a friend chide me that I paid too much. As it would turn out, I got the bargain of a lifetime. When I researched the historical background of the coin, I realized that I held in my hand a link to the story about the Star of Bethlehem. I felt then compelled to write an article about that coin to raise interest in further research. That article resulted in an important telephone call, which would lead to writing this book.

Prof. Owen Gingerich of Harvard University had read my 1992 article, *The Coins of Antioch*, in Sky & Telescope magazine and called me at Rutgers University to say that he too had noticed these Roman coins when he was at the American University of Beirut in Lebanon. Gingerich said people found the coins in the dust along ancient roadways, but he never found time to analyze their meaning. Gingerich was excited that I had shown that Quirinius, the Roman governor of Syria, issued these coins. In the Book of Luke about the birth of Jesus, Roman Emperor Augustus Caesar ordered Quirinius to seize control of Judea to suppress unrest fueled by rumors of a newborn King of the Jews, the Messiah. A well-known prophecy claimed that the Messiah would lead the "forces of light against the dark", namely pious Jews against hated pagan Romans.

As I explained in that seminal article, the coins appear to be a Roman response to messianic rumors. Romans avidly practiced astrology and solidly embraced it to guide personal and even political decisions. For instance, during the timeframe of the Star of Bethlehem Roman Emperor Augustus Caesar had a court astrologer, Thrasyllus, for advice in astrological matters. The coin issued by Quirinius depicted Aries the Ram that was the zodiacal sign for Judea according to many ancient astrological sources. I concluded that the coin showed where the biblical star would have appeared in the sky. The question facing me was what happened in Aries the Ram to signify the birth of a King of the Jews.

Based on extensive research into astrological predictions of Roman times, I learned that a king's birth or his coronation was often associated with a very close conjunction between the planet of kings, namely Jupiter, and the life-giving Moon. For example, when Octavian became Roman Emperor Augustus Caesar on January 16, 27 BC, the Moon occulted (passed in front of) Jupiter twice in the following days in the sky over Rome. Augustus Caesar had issued a coin at this time depicting a star and crescent, a symbol of the Moon occulting a planet. As I show in this book, I had found numerous examples of Roman Emperors tying their imperial fate to this kind of astrological event. Surely, the Star of Bethlehem involving this sort of regal portent would have had the rapt attention of Romans.

My calculations pointed to two possible dates when astrologers would have claimed that the King of the Jews was born. I wrote that I had no idea which date, March 20, 6BC or April 17, 6BC, was correct. Believing at first that this article concluded any further involvement, I changed my mind after listening to Gingerich's collegial haranguing. He explained that I was close to the answer, and should not disband future work. He was right!

A few months later, recalling Gingerich's remarks, I examined the original Greek text of the Book of Matthew that describes the visit to King Herod by the Magi—stargazers adept at explaining planetary alignments. They proclaimed to King Herod and Jews of Jerusalem that a star announcing the birth of the King of the Jews was "en te anatole". I recognized this Greek phrase from ancient astrological texts. The Magi undoubtedly spoke of a morning star, a messenger of God for Jews. Realizing the true meaning of this ancient astrological term, I rushed to my computer to recreate this celestial condition. Incredibly, I found Jupiter "in the east" precisely on the second of the two candidate dates, April 17, 6BC, cited in my Sky & Telescope article. Now, I was very interested in the Star of Bethlehem and felt compelled to write this book.

Writing this book helped organize my thoughts and focus my research. I uncovered even more historical evidence supporting an actual historical basis for the Star of Bethlehem. For instance, I realized that readers would ask whether I ever came across any reference to the biblical star in all of the extensive ancient documents I had collected. Thanks to the incredible library services of Rutgers University, I had amassed a treasure trove of ancient resources. As it turned out there was indeed such a reference sitting among those books.

The Roman Christian convert, Julius Firmicus Maternus, wrote a book about astrology during the time of Emperor Constantine the Great (ca. AD 346). Firmicus also wrote a book against paganism because fellow Christians challenged his faith for writing a book on astrology that included many pagan references. In

his astrology book, The Mathesis, Firmicus described the astrological conditions for the birth of a divine and immortal person.
Bound by Roman laws for astrologers, Firmicus could not reveal
names when interpreting astrological conditions, but he did in
fact describe conditions matching precisely what happened in
Aries the Ram on April 17, 6BC. He wrote that the person born
on this day was "surely divine and immortal". His report, I argue
in this book, describes the birth of Jesus. This supports my finding that, stargazers had seen a sign of the birth of the King of
the Jews two years before King Herod died. Jews keeping vigil
for the Messiah would have interpreted this as a message from
God that deliverance from the Roman yoke was at hand. This
is probably the reason why the Jewish historian, Josephus, said
that before King Herod died there were rumors of the Messiah.

Following the publication of this book, Michael Hoskin, editor
of the Journal for the History of Astronomy, invited me to write
a scientific paper challenging other astronomers and historians
to disprove the fundamental basis for my explanation of the Star
of Bethlehem. Namely, my findings are that the zodiacal sign of
Aries the Ram was the ancient astrological sign of the Jews and
this is where the Star of Bethlehem appeared. If my idea about
Aries the Ram is wrong, so is my entire explanation. Other astronomers have advocated different signs of the zodiac such as
Pisces the Fishes that many people of faith favor because the
fish is symbolic of Christianity. However, the historical evidence
is that it was up to neither Christians nor modern astronomers to
define the sign of the Jews. According to my extensive research,
astrologers had made this assignment for Judea to Aries the Ram
before Christianity existed. In any case, my challenge appears in
The Evidence for Aries the Ram as the Astrological Sign of Judea
(JHA, vol. XXXIV, p.325, 2003). To date, no one has responded
because the amount of supportive evidence I have amassed is
daunting for anyone advocating a pet theory.

The reaction to this book has been outstanding. There are
now Italian and Czech versions. Countless news organizations
interviewed me and I flew to the United Kingdom to record a
television program for the BBC. I also presented many seminars,

but one at NASA Goddard Space Flight Center was particularly memorable. I had worked there in 1971 on the Orbiting Astronomical Observatory-2 satellite project. The audience was enthusiastic with people standing in the aisles, but most importantly to me, my former graduate students from The University of Toledo also attended. The circuit of seminars took me to Harvard University where Professor Gingerich showed me his coins of Antioch with Aries the Ram. Years later, I taped a show about the Star of Bethlehem with renowned Egyptologist Bob Brier. That program and others appear repeatedly each holiday, which is a fitting tribute to the success of this book.

Some reactions have not been positive. I was amazed to find erroneous reports concluding that I gave astrology credence in Christianity. Nothing could be farther from the truth. I am a scientist who adamantly rejects the notion that planetary configurations control people as astrologers claim nowadays. Historians tell us that during biblical times, Jews looked to the stars for messages from God. In this book, I show passages in the Old Testament and from other sources, how Jews believed that the morning stars, namely bright planets, were angels—heralds of God. Jews would have listened to the prognostications of stargazers such as the Magi to help interpret the mysterious planetary motions. Nevertheless, Jews did not practice astrology as their pagan neighbors did.

Most people of faith have been fair and open minded about this book. Churches and religious groups invited me to speak. Even astrologers, skeptical of an astronomer, asked me to present my case, which they received well. However, there was an unexpected reaction from skeptics of anything involving the Bible. On two occasions, I found commentaries on this book that were very contrary to what I had written. After I questioned the writers, it became apparent that they did not read this book. I received apologies, but to no help of my damaged reputation. I hope that people understand how such intellectual dishonesty is truly counterproductive and poisonous to uncovering the truth.

It is understandable how topics related to religious beliefs can raise emotions, and put people on guard when hearing about this book. Strong religious beliefs can spur people to alter facts in order to promote their personal convictions that may even lie beyond what evidence can support. For example, some other theories of the Star of Bethlehem discussed in this book alter historical dates in order to achieve better agreement between the two Nativity Narratives of Luke and Matthew. Even the famous German astronomer Johannes Kepler writing in the early 1600's took liberty with the facts in his explanation of the Magi's star. As I show in this book, Kepler did not apply Roman-era concepts, but used Renaissance ideas. Moreover, we know that he formed his ideas through his powerful Protestant faith and mystical beliefs.

I want to state categorically that I present these findings truthfully as an unbiased scientist. In establishing dates and astrological practices, I use reputable sources backed by solid scholarship. I challenge anyone to find any anti-religion bias or a pro-Christian agenda in this book. Furthermore, I have been scrupulously honest with historiography, the methodology for interpreting history. Researchers examining historical reports such as the Star of Bethlehem can only use sources from a similar time and culture. This means that I used only documents relevant to Jewish-Roman times and traditions. For example, I do not use Mayan or Chinese astrological reports that stem from irrelevant cultures, or Renaissance Christian concepts that come from the wrong culture and era. Too many researchers have not followed these rules producing only confusion and perhaps cynicism about any possible historical basis to the Star of Bethlehem.

I trust that readers can see how each piece of evidence I uncovered changed my own reluctance to become involved with the Star of Bethlehem. What once I thought to be a pious myth, I now know has a valid historical basis. Nevertheless, I cannot prove that Jesus was born under this star. *That* is indeed a matter of faith!

Michael R. Molnar, Ph.D.
Warren, NJ

*

The Star of
Bethlehem

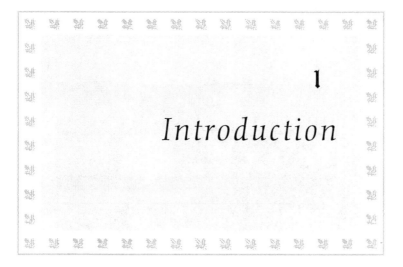

1

Introduction

A Sign from a Coin

When winter nights grow long, our lives are brightened by the celebration of Christmas. This time of peace and good will is celebrated throughout the world as many people commemorate the birth of Jesus, which was marked by a special celestial manifestation, the Star of Bethlehem. A star suspended above a crèche reminds us of the biblical account in Matthew of the Magi, who came to King Herod of Judea to ask where the newborn King of the Jews was, for they had "seen his star in the east" (see figure 1).

The celestial event that is believed to have revealed the birth of Jesus has captivated the imagination of many people over the centuries. Thus, there are many explanations as to what the "star" might have been: a comet, a supernova, a planetary conjunction, or even a miracle. The Star of Bethlehem is mentioned only in Matthew; no contemporaneous historical source mentions the

Figure 1. The Nativity, a woodcut by Albrecht Dürer, from "The Little Passion" (1508–1510), typifies crèches that display the Star of Bethlehem as a single star above the infant Jesus.

star.[1] Furthermore, much lore has been added to the biblical account, and strong personal beliefs have embellished the facts, making it difficult to determine what the Magi saw in the sky.

The account in Matthew is somewhat enigmatic. Visitors come to King Herod, and they are called Magi, a title that bestows an aura of mystery. The Magi are from somewhere in the East, and they bring valuable gifts for a newborn king. Their eastern origins and the precious presents conjure up visions of exotic places and great wealth, and even suggest to some people that the Magi were kings. Their star is also mysterious. It is difficult to visualize how the Magi, coming from the east, could have followed a star "in the east" to Judea. Furthermore, perhaps miraculously, the star seems to stand over where the child was. And most mysterious of all, it appears that no one in Jerusalem saw the star. All of this information points to a rather unusual celestial event, one as perplexing to us as it was to King Herod and his people.

In this book, I will present new evidence in favor of a historical basis to the star mentioned in Matthew. There was indeed a great celestial portent during Herod's reign, a portent that signified the birth of a great king of Judea and is in excellent agreement with the biblical account. An interesting feature of this new evidence is its serendipitous nature—its fortuitous emergence from studies of ancient coins.[2]

Late in the reign of Emperor Augustus Caesar, close to A.D. 6, Antioch, the capital of the Roman province of Syria, issued some bronze coins of low denominational value for local use. Although Antioch had minted coins many times before, these new coins were remarkably different from previous issues. They displayed a leaping ram, the zodiacal sign of Aries, looking backwards at an overhead star.[3]

The appearance of an astrological symbol on a Roman coin is not unusual: astrology was widely practiced throughout the Roman world, even at the highest levels of society and government. For instance, Emperor Augustus Caesar issued several coins bearing his birth sign, Capricorn, the Sea-Goat. Astrologers claimed that Augustus's astrological birth chart predicted he would rule

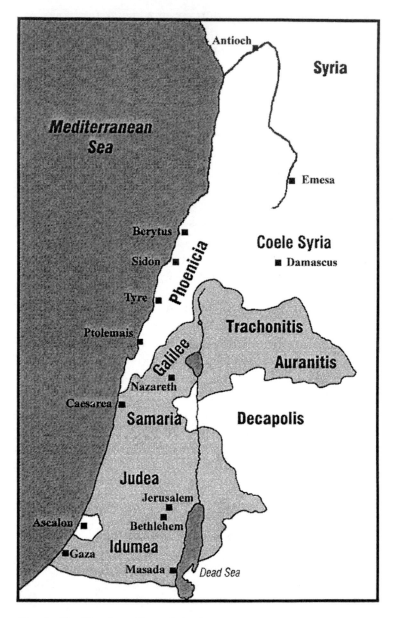

Map 1. The Kingdom of Herod the Great

the world, so Augustus used Capricorn as propaganda—proof from the heavens that he was destined to become emperor. As a result, Capricorn became a common emblem on coinage and artwork across the empire.[4]

As for the meaning of Aries, the Ram, on Antioch's coins, numismatic scholars have surmised that Aries was Antioch's astrological sign—that the city was "born" under this sign.[5] That explanation is not, however, entirely correct: the ancient astrological texts report that Aries also symbolized Judea (see map 1 for the location of Antioch relative to Judea). Thus, the coins may have been related to the annexation of Judea by the Romans in A.D. 6. But an even more important possibility, one supported by astrological sources, is that the coins explain that Aries is where stargazers would have watched for indications of the birth of a king of the Jews. This means that the Star of Bethlehem would have appeared in Aries. But before asking what actually did appear in Aries, we must examine the historical background of the story about the Star of Bethlehem.

THE ORIGINS OF THE BIBLICAL ACCOUNT

The stories about the birth of Jesus, which are known as the infancy narratives, come from two sources, the Gospels of Matthew and Luke in the New Testament. It is likely that Matthew's story of the Magi and their regal star is not a firsthand account but rather the product of an oral tradition that was eventually recorded long after Jesus' ministry. Most biblical scholars are of the opinion that Matthew and Luke were written in the period A.D. 80–90, fifty to sixty years after the crucifixion of Jesus. The author of Matthew is unknown, although the evidence points to a Greek-speaking Jewish Christian who lived among Gentiles and Jews in Syria, most likely in Antioch. Similarly, the evangelist of Luke is unknown but believed to have been a physician who lived in Antioch and wrote the best Greek of the evangelists. Furthermore, he appears to have been a companion of Paul the Apostle. Both Gospels indicate that some material was drawn from the Book of Mark, the earliest gospel, and from another source of Jesus' sayings called "Q" by biblical scholars.[6]

There are strongly differing views about the account of the Magi's visit. One view holds that the story about the regal star is purely a myth meant to give the birth of Jesus a glorious and mystical setting. Another interpretation is that the star defies explanation and was a miracle. Between these diametrically opposed beliefs lie many explanations that try to tie the star to a specific celestial event. These theories about the star are strongly shaped by personal ideas about celestial portents. And above all, the people who present these theories are unfamiliar with the beliefs from the time of Jesus' birth about such portents.

Some researchers have argued that the tale about the mysterious star is midrash, a form of ancient Jewish interpretation and exposition that popularizes and explains a biblical account.[7] Some scholars argue that the story of the star, although in the literary style of midrash, is nevertheless true; and others argue that all midrash is a form of myth, that the star was conjured up to validate Jesus' birth as Christ (the Messiah.)

The argument advocating a mythical account points to the prophetic oracle of Balaam from the Old Testament, which indicates that Jews expected the appearance of a Messianic star that would mark the birth of the Messiah.

> there shall come a Star out of Jacob, and a Sceptre shall rise out of Israel, and shall smite the corners of Moab, and destroy all the children of Sheth. (Num. 24:17)

Balaam, a famous seer, was summoned by King Balak, the Transjordanian king of Moab, to put a curse on the Israelites, who were being led by Moses out of Egypt to find the Promised Land.[8] Balaam thwarted the plan of King Balak by foretelling not the downfall of Israel but its future greatness. Some scholars, pointing to versions of the text in which "star" is replaced by "king," believe that this passage predicts the rise of King David. In any case, the story was used for centuries as a prophecy about the Messiah, who would destroy the enemies of Israel. Moreover, many parallels between the story of Balaam and the Magi's visit have been noted: King Balak was from the same land as King Herod's family; Balaam ruined Balak's plans to destroy the Israelites, and the Magi foiled Herod's plan to destroy Jesus; Balaam, like the Wise Men, was a *magos* (a seer); and Balaam spoke of a star symbolizing the Messiah, and the Magi said a star announced the birth of the King of the Jews, namely, the Messiah.

These and other similarities between the stories have been used to argue that the account in Matthew is a midrash fable based upon the story of Balaam and that the account in Matthew needed to invent a Messianic star to establish Jesus as the Messiah. Therefore, the argument goes, if the Messianic star of Matthew cannot be verified as an actual celestial event, midrash, based upon the story of Balaam, is the explanation of the star.[9] In view of the numerous and unconvincing explanations that try to establish a historical basis for the Magi's star, it is not surprising that the story about the star is assumed to be an intriguing myth or even a report about a supernatural apparition. For these reasons some people believe that the enigmatic star can be accepted only as a matter of faith.

Nevertheless, many people suspect that there is a yet-to-be-found historical basis to the intriguing story. The details in the account of the Magi's visit strongly suggest an attempt, albeit a confusing one, to pass along details of an important celestial manifestation—one that was noted or even anticipated by people watching for celestial omens about the advent of the Messiah. For this reason many researchers have described the story as verisimilar, that is, likely, or probable. However, the problem is that a verisimilar account is not proof for a historical basis.

Another popular opinion is that the star was a real celestial phenomenon but that its enigmatic nature is due largely to its conveyance by people who did not understand the meaning of the Magi's message. That is, the evangelist's recording of the arcane terminology used by the Magi obscured the description of their star, and for that reason the account's true historical basis has remained hidden. Thus, the star appears to be a myth only because astronomers have failed to identify an astral portent that unmistakably points to the birth of a king of Judea.

THE JEWS IN THE ROMAN WORLD

During King Herod's reign there were growing expectations of the advent of the Messiah: people were watching for a sign or an omen of a great ruler who would rise up from Judea and vanquish the country's enemies, namely, the tyrannical Romans and their puppet king, Herod. Therefore, we must first examine the historical background for the anticipated deliverance by the Messiah to appreciate fully the significance of the Magi's visit.

The relevant events leading up to the visit of the Magi start essentially with Rome's efforts to control Judea, efforts that raised Messianic fervor there. The small country was strategically significant for Roman plans to dominate the eastern Mediterranean because it was on the route between Egypt, an important source of Rome's food supply, and the rich trade routes of Mesopotamia, where the feared Parthians ruled (see map 2). Thus, Judea found itself drawn into many struggles in which the stakes were high.

A couple of years after the victory (42 B.C.) of Octavian (the future Augustus Caesar) and Mark Antony over the assassins of Julius Caesar, events dramatically unfolded for Judea. To the east, in Mesopotamia and Babylonia, was the powerful Parthian kingdom, which kept watch on Roman conquests. Moving to check Roman expansion in the Near East, the Parthians invaded Syria and accepted an invitation of some Jews to help overthrow the Roman pawn (Hyrcanus) ruling Judea. Herod, a Roman ally, fled Judea to Egypt, and Cleopatra helped him to sail to Rome. Octavian and Mark Antony threw their support behind him, making him king of Judea in 40 B.C.

Herod quickly returned to Judea with Roman legions to take back his kingdom from the Parthians and rebellious Jews. However, he found himself in the awkward position of attacking his own countrymen with foreign forces. Although Herod stopped

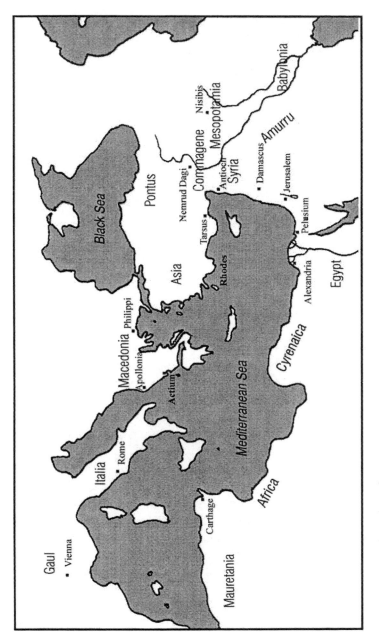

Map 2. *Historical Lands and Cities*

the Romans from completely pillaging Jerusalem, he was disliked, if not hated, by his subjects, who saw him as another puppet placed on the Judean throne by the despised Romans. Herod finally took control of Judea in 37 B.C. and began a ruthless reign in which he moved quickly to snuff out any threats, real or imagined.

Herod embarked on an ambitious building program for his kingdom. In 20 B.C. he initiated the construction of Caesarea, which became a showcase city in the Near East. He also started many projects that benefited the economy, including a major renovation of the Temple of Solomon, which was not finished until well after his death.[10] Although Herod succeeded in statesmanship and economic achievements for Judea, he failed miserably in securing the support and loyalty of his people. Herod was not from the royal Hasmonean family, and still worse, he was not seen as a Jew. His mother was a Nabatean, and the rest of his family was from Idumea, where the people had been forced to convert to Judaism. Furthermore, Herod had been raised not as a Jew but rather in Hellenistic traditions. Although Herod did keep a kosher table in the company of Jews, he avidly pursued cosmopolitan programs outside Jerusalem, engaging, for example, in the construction of pagan temples and projects in Caesarea, Antioch, Tyre, and elsewhere.[11] His interest in Greek culture led to his support for reestablishing the Olympic Games in Greece. Thus, Jews viewed him as a Hellenistic tyrant forced upon them by the Romans.

It was Herod's brutal authoritarianism, above all, that distanced him from his people. He quickly and ruthlessly suppressed any perceived threat to his authority: Herod killed three of his sons and threatened to kill the rest of his children. Commenting on the slaughter of Herod's sons, Augustus Caesar astutely observed that it was safer to be Herod's pig than Herod's son.[12]

Many Jews saw these as apocalyptic times. Roman tyranny, compounded by the presence of a ruthless "outsider" on the Judean throne, proved to be unbearable. Thus, Jews turned to their faith for salvation. Within the Jewish community were active fundamentalists who preached that a messiah would rid Judea of its

Roman enemy and puppet king. Such Messianic dreams grew
rapidly during the reign of Herod, and they would persist for well
over a century and produce two major revolts. One group of Messianic believers, the Qumran community, produced the famous
Dead Sea Scrolls. Among these ancient documents is the *War
Scroll,* which tells about a great battle that will come in the end
between the faithful of Qumran and the Hellenistic peoples: the
Sons of Light versus the Sons of Darkness.[13]

Herod had his hands full indeed and reigned as a lonely monarch who was reactive to Roman demands, watchful for Parthian
encroachments, doubtful of his family's loyalty, and suspicious of
his Jewish subjects, who were waiting for the Messiah. It is within
this historical context that the biblical account about the Magi
must be considered.

The Account in Matthew

The account of the Magi's visit to King Herod appears in Matt. 2:
1–16. Table 1.1 gives the King James Version, with which most
people are familiar, and the New Revised Standard Version,
which is a modern English interpretation. There are several features of the account that are important to any interpretation of
the Star of Bethlehem.[14] The Magi did not inquire whether a king
had been born. Rather, they knew according to their prognostications that the young prince was somewhere in Judea. They
assumed that Herod too knew about the new king, but they found
instead that their emphatic pronouncement stunned him. The
biblical account reports that Herod's advisors and the people of
Jerusalem were surprised and puzzled by the Magi's report of a
regal star. Apparently, the Magi's star was either invisible or, more
likely, a subtle manifestation that passed unnoticed by people
who did not understand the practices of the Magi. Having been
raised in Greek traditions, Herod appreciated the meaning of
the Magi's message, and his reaction reveals something fundamentally important, namely, that he believed that they were talking about the Messiah, not just any king. Because the Messianic
prophecy that a king of the Jews would conquer the world was

Table 1.1. Matthew 2: 1–16

KING JAMES VERSION	NEW REVISED STANDARD VERSION
Now when Jesus was born in Bethlehem of Judæa in the days of Herod the king, behold there came wise men from the East to Jerusalem,	In the time of King Herod, after Jesus was born in Bethlehem of Judea, wise men from the East came to Jerusalem,
Saying, Where is he that is born King of the Jews? for we have seen his star in the east, and have come to worship him.	asking, "Where is the child who has been born king of the Jews? For we observed his star at its rising, and have come to pay him homage."
When Herod the king had heard *these things,* he was troubled, and all Jerusalem with him.	When King Herod heard this, he was frightened, and all Jerusalem with him;
And when he had gathered all the chief priests and scribes of the people together, he demanded of them where Christ should be born.	and calling together all the chief priests and scribes of the people, he inquired of them where the Messiah was to be born.
And they said unto him, In Bethlehem of Judæa: for thus it is written by the prophet,	They told him, "In Bethlehem of Judea, for so it has been written by the prophet:
And thou Bethlehem, *in* the land of Juda, art not the least among the princes of Juda: for out of thee shall come a Governor, that shall rule my people Israel.	'And you Bethlehem, in the land of Judah; are by no means least among the rulers of Judah; for from you shall come a ruler who is to shepherd my people Israel.'"
Then Herod, when he had privily called the wise men, enquired of them diligently what time the star appeared.	Then Herod secretly called for the wise men and learned from them the exact time when the star had appeared.
And he sent them to Bethlehem, and said, Go and search diligently for the young child; and when ye have found *him,* bring me word again, that I may come and worship him also.	Then he sent them to Bethlehem, saying, "Go and search diligently for the child; and when you have found him, bring me word so that I may also go and pay him homage."
When they had heard the king, they departed; and, lo, the star, which they saw in the east, went before them, till it came and stood over where the young child was.	When they had heard the king, they set out; and there, ahead of them, went the star that they had seen at its rising, until it stopped over the place where the child was.

KING JAMES VERSION	NEW REVISED STANDARD VERSION
When they saw the star, they rejoiced with exceeding great joy.	When they saw that the star had stopped, they were overwhelmed with joy.
And when they were come into the house, they saw the young child with Mary his mother, and fell down, and worshipped him: and when they had opened their treasures, they presented unto him gifts; gold, and frankincense, and myrrh.	On entering the house, they saw the child with Mary his mother; and they knelt down and paid him homage. Then, opening their treasure chests, they offered him gifts of gold, frankincense, and myrrh.
And being warned of God in a dream that they should not return to Herod, they departed into their own country another way.	And having been warned in a dream not to return to Herod, they left for their own country by another road.
And when they were departed, behold, the angel of the Lord appeareth to Joseph in a dream, saying, Arise, and take the young child and his mother, and flee into Egypt, and be thou there until I bring thee word: for Herod will seek the young child to destroy him.	Now after they left, an angel of the Lord appeared to Joseph in a dream and said, "Get up, take the child and his mother, and flee to Egypt, and remain there until I tell you; for Herod is about to search for the child, to destroy him."
When he arose, he took the young child and his mother by night, and departed into Egypt: And was there until the death of Herod: that it might be fulfilled which was spoken of the Lord by the prophet, saying, Out of Egypt have I called my son.	Then Joseph got up, took the child and his mother by night, and went to Egypt, and remained there until the death of Herod. This was to fulfill what had been spoken by the Lord through the prophet, "Out of Egypt I have called my son."
Then Herod, when he saw that he was mocked of the wise men, was exceeding wroth, and sent forth, and slew all the children that were in Bethlehem, and in all the coasts thereof, from two years old and under, according to the time which he had diligently enquired of the wise men.	When Herod saw that he had been tricked by the wise men, he was infuriated, and he sent and killed all the children in and around Bethlehem who were two years old and under, according to the time that he had learned from the wise men.

SOURCE: *John R. Kohlenberger III, ed.,* The Precise Parallel New Testament *(New York: Oxford University Press, 1995), 6–9.*

well known,[15] it is evident that a visit by people looking for a new-born king—perhaps the Messiah—would have been alarming for Herod.

Herod "diligently" quizzed the Magi about the star, making special note of when it had appeared so that he could gauge the age of the child. His advisors said that it was prophesied that the Messiah would be born from the lineage of King David. The home of David was Bethlehem, so Herod sent the Magi there. Obviously planning to strike at this threat to his throne, he asked them to report back to him after finding the child. According to Matthew, the Magi found the infant Jesus in Bethlehem; thus, the Magi's star became known as the Star of Bethlehem. It is odd that the Magi told a king about the birth of another king without expecting serious repercussions. Nevertheless, the Magi eventually recognized Herod's duplicity and secretly returned to their country without reporting back to him.

Two important pieces of the puzzle of the Magi's star lie in determining how the Magi knew (1) that a king had been born and (2) that he had been born in Judea. Why did they look to the sky for indications of a regal birth, and what happened in the sky that signaled such a birth?

2

A Mysterious Star

The Magi's star has been the subject of innumerable studies and interpretations, and sometimes scholars involved in these studies have taken liberties with the historical background to justify a popular or, more often, personal notion. Nevertheless, we can learn much by analyzing these interpretations.

The important connection between the Star of Bethlehem and powerful religious beliefs about Jesus has moved some people to advocate a supernatural apparition as an explanation for the star. Miracle theories usually point out that the unusual behavior of the Magi's star (it "went before them" and "stood over where the young child was") seemingly defies a natural explanation and that, therefore, the star must be a supernatural manifestation. Often these theories suggest that the star was an angel because, according to Jewish and Christian beliefs, an angel is a messenger of God, serving as an interface between God and humans; and there are biblical references in which stars are seen as messengers of God.[1]

> When the morning stars sang together, and all the sons of God shouted for joy? (Job 38:7)

> He telleth the numbers of the stars; he calleth them all by their names. (Ps. 147:4)

> The heavens declare the glory of God; and the firmament sheweth his handiwork. (Ps. 19:1)

Nevertheless, it is difficult to accept a miracle or an angelic messenger as an explanation for the star unless all possible natural explanations have been refuted, and there are several reasons for seeking a natural explanation. Many researchers note that Matthew does not call the star a miracle, nor does the account even suggest that the Magi were talking about a supernatural apparition. Most important, the remarkably good fit between the story about the Magi's visit to Herod and the political and social context tends to support a natural, physical explanation:

- Astrology was widespread throughout the Roman world, especially in the Near East, and practitioners of astrology were highly respected. The Magi would naturally have been permitted to have an audience with a king.
- The Jews were anticipating the arrival of a messiah who would deliver them from Roman tyranny. The prophecy from the story of Balaam would have suggested to many Jews the appearance of a Messianic star.

Although Jews did not practice astrology, Hellenistic people did and would have looked to astrology for indications of the fulfillment of the Messianic prophecy. Any viable explanation of the account in Matthew cannot ignore that historical context and must be evaluated according to the beliefs of the people of Herod's time, not according to the notions of modern people.

If the theory that the star is an unexplainable miracle is put aside, there are several intriguing theories that relate the star to a specific celestial event. And often these theories strongly incline toward advocating astronomical phenomena, that is, visible movement or positioning of celestial bodies, as portents. An astronomical event by itself does not necessarily have underlying meaning or significance; however, theorists sometimes use statistical calculations to show that an event was rare, and its rarity leads them to believe that the event must have been important for astrologers. These popular theories draw conclusions about astrological portents with little regard to the record of ancient beliefs.

All of the following theories of the Star of Bethlehem are based upon an interpretation of the Greek word for star (ἀστήρ = *aster*) used in the biblical account. Matthew does not explain what kind of star the Magi recognized, and the analysis of the story is complicated by Greek writers' use of the term *aster* to describe several different kinds of celestial phenomena, including comets, also known as "long-haired stars" (κομῆται = *cometai*).

The Comet

That the Star of Bethlehem was a comet is one of the most popular theories, and its popularity is understandable because comets are sensational. A comet is essentially a huge icy asteroid. The majority of comets are in highly elliptical solar orbits, and they spend most of their time at great distances from the Sun, traveling slowly through the deep freeze of space. Comets remain undetected until they come swooping in close to the Sun. Then the solar heat vaporizes the ice and produces magnificent, gossamer tails that stretch far across the sky. These unexpected interlopers are visible for weeks and even months, slowly wandering across the sky. Some historians have suggested that a comet literally pointed the way to Judea for the Magi, and several recorded comets have been cited as potential candidates for the Magi's star.[2]

The evidence, however, is strong that people of Roman times feared rather than welcomed comets. Long-haired stars were thought to be harbingers of disaster, usually the death of a king or an emperor. Roman biographer Suetonius gives us an excellent example of the fear of comets: when one appeared during the reign of Vespasian in A.D. 79, the cagey emperor knew that his demise was on the mind of his colleagues, especially opportunists hoping to exploit the perceived danger.

> He did not cease his jokes even in apprehension of death and in extreme danger; for when among other portents . . . a comet appeared in the heavens, he declared that [it] applied . . . to the king of the Parthians, who wore his hair long. (*Vespasian* 23)

There is a wonderful pun in this story: the Parthian kings were known for wearing long hair (κομήτης = *cometes* = wearing long hair), whereas Vespasian was bald. In other words the long-haired star of death was meant for the long-haired monarch. Vespasian had indeed a tremendous sense of humor. Nevertheless, people probably thought the joke was on him—he died that year, and his death undoubtedly confirmed everyone's fears about the evil nature of comets.

Researchers advocating a comet as the Magi's star point to instances in history when people questioned the purported evil nature of comets or when historical notables interpreted a comet as a blessing. One of the most celebrated comets appeared after the assassination of Julius Caesar in 44 B.C. Augustus Caesar, Julius Caesar's adopted son and political heir, knew that people would speculate that the new comet foretold his own death. Moreover, such rumors often became self-fulfilling prophecies; that is, enemies would strike believing their adversary was weakened by the omen.[3] However, Augustus stemmed any thoughts about his demise by proclaiming that the comet was the wandering soul of Julius Caesar. Augustus proved to be one of history's greatest propagandists and spin-control artists: he commissioned coins and statues honoring the comet (see figure 2), which made historian Pliny the Elder (A.D. 23/4–79) remark incredulously:

> The only place in the whole world where a comet is the object of worship is in a temple at Rome. (*Natural History* 2.92)

*Figure 2.   This silver denarius issued (19–18 B.C.)
by Augustus Caesar shows the comet of "Divine
Julius," which was cleverly propagandized as a
good portent. Author's collection, RIC-1 37a*

Another person who tried to put a new spin on the meaning
of comets was the Roman senator Seneca, who served as Nero's
tutor. That Seneca wrote about the good nature of comets in his
Natural Questions is often cited as evidence that comets could
herald the birth of a king. During Nero's reign the appearance of
a comet in A.D. 60 made Romans think about Nero's replacement:

> Meanwhile, a comet blazed into view—in the opinion of the
> crowd, an apparition boding change to monarchies. Hence, as
> though Nero were already dethroned, men began to inquire on
> whom the next choice should fall. (Tacitus *Annals* 14.22)

Seneca tried to put a positive image on the political and social
events of his time by citing how good things were under Nero (!)
and by proposing that the comet was responsible for this good
fortune. Tacitus, however, reported that a second comet in A.D. 64
made Seneca eat his words.[4] That year was truly an *annus hor-
ribilis:* a great fire swept through most of Rome, and major

conspiracies against Nero were exposed. Furthermore, Nero's insensitive comment that the fire enabled him to build a better city caused outrage and eventually produced the famous myth about how he "fiddled" while Rome burned. Nero tried to deflect the public's outrage by blaming the fire on the Christians. And Suetonius tells us that the astrologer Balbillus devised a way to protect Nero from the comet:

> It chanced that a comet had begun to appear on several successive nights, a thing which is commonly believed to portend the death of great rulers. Worried by this, and learning from the astrologer Balbillus that kings usually averted such omens by the death of some distinguished man, thus turning them from themselves upon the heads of the nobles, he resolved on the death of all eminent men of the State; but the more firmly, and with some semblance of justice, after the discovery of two conspiracies. (*Nero* 36)

Seneca's upbeat opinion about comets fell on deaf ears: Nero forced Seneca and many other prominent people to choose between committing suicide and being killed as conspirators. There is evidence that several leading citizens went to their death as surrogate victims to placate the bloodthirsty comet.[5]

The idea that comets could signal the birth of a new king is unsupported by the reports of contemporaneous historians. Table 2.1 gives a list of cometary interpretations compiled by astronomy historian and classicist A. A. Barrett, a list which spans the years close to the birth of Jesus. According to these sources comets were commonly believed to signify death and battles, not royal births.[6]

Astrologers' perceptions about comets can be summed up by a quotation (ca. A.D. 150) from Claudius Ptolemy, an authority on Greek astrology:

> For these [comets] naturally produce the effects peculiar to Mars and to Mercury—wars, hot weather, disturbed conditions, and the

Table 2.1. Cometary Portents Close to the Time of Jesus' Birth

Year	Interpretation of comet recorded by Roman historians	Source
42 B.C.	Battle of Philippi: Augustus and Antony defeat Brutus and Cassius.	Dio Cassius 47.40.2; Manilius 1.907; Virgil, *Georgics* 1.488
31 B.C.	Battle of Actium: Augustus and Agrippa defeat Antony and Cleopatra.	Dio Cassius 50.8.2
30 B.C.	Death of Cleopatra.	Dio Cassius 51.17.5
17 B.C.	Fighting breaks out in Gaul; unrest in Rome.	Dio Cassius 54.19.7; Obsequens 71
12 B.C.	Death of Marcus Agrippa.	Dio Cassius 54.29.8
A.D. 9	Three Roman legions under Varus are annihilated by the Germans	Manilius 1.899; Dio Cassius 56.24.4
A.D. 14	Death of Augustus Caesar.	Dio Cassius 56.29.3

accompaniments of these; and they show, through the parts of the zodiac in which their heads appear and through the directions in which the shapes of their tails point, the regions upon which the misfortunes impend. (*Tetrabiblos* 2.9)

Thus, a comet pointing to the sign of Judea would not have meant good news about a birth of a king of the Jews. Of course, it could be argued that the death of a king would lead to a new king. Nevertheless, the account in Matthew tells us that the Magi were searching for a newborn king, not a doomed king.

Finally, a comet would have been easily noticed by Herod and the people of Jerusalem; yet they did not see the star. Their failure to see the star, along with the fear of comets, shows that the Magi's regal star could not have been a comet.

The Supernova

Sometimes a star suddenly appears in the sky as if by magic and remains for a few weeks. This dramatic and spectacular apparition is called a *nova,* a "new star." Astronomers now know that a nova is not a newborn star but a dying star or one in the late stages of its life. A nova occurs in a special binary star system in which a swelling, aging star pours gaseous material onto its orbiting neighbor, a small hot star called a white dwarf. The accumulated material flares in a runaway nuclear explosion on the surface of the white dwarf, producing a nova.

There is an even more luminous occurrence called a *super-nova,* of which there are two types.[7] In the first, an aging, huge star completely consumes its nuclear material and can no longer generate enough energy to support its own weight. The star collapses and produces a violent explosion. The second type occurs when a white dwarf accumulates too much matter from a neighboring star. If the white dwarf cannot shed the extra matter (as in a nova), it collapses, producing a violent supernova explosion.

In both of these cases a brilliant star suddenly appears in the sky and slowly fades back into obscurity over a period of several months. Of course, the stargazers of antiquity had no idea that these "new stars" were in fact doomed stars. But most people today visualize the Star of Bethlehem as a single bright star, and the supernova theory beautifully fits that notion.

The theory that the Magi's star was a nova or supernova owes its origin primarily to the great astronomer (and astrologer) Johannes Kepler, who studied a new star that appeared on October 10, 1604, in the constellation Ophiuchus. By chance, this stellar apparition appeared among a triple conjunction of Mars, Jupiter, and Saturn that had started in September in the zodiacal sign of Sagittarius (see figure 3). At the time there were speculations that great planetary conjunctions would produce a bevy of comets, which, as we have seen, people feared as portents of death and destruction. The idea that conjunctions produced comets was widely held by medieval followers of Aristotle, the great Greek philosopher, who taught that a comet was the result of light emanating from planets during a conjunction. Kepler took a strong interest in close conjunctions to verify this theory. Because conjunctions usually involved only two planets, not three, there was much apprehension as to whether the triple conjunction of 1604 would produce any ominous comets.[8]

Much to everyone's relief, no evil comet appeared during the triple conjunction. Instead, just as Mars ♂ was leaving the group,

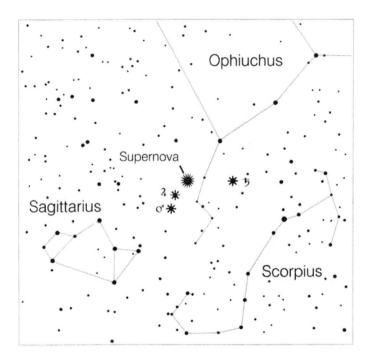

*Figure 3. The appearance of Kepler's supernova among the plan-
ets of the triple conjunction in 1604 gave the impression that the
planetary gathering brought forth the new star.*

a supernova appeared between Jupiter ♃ and Saturn ♄. The
bright new star was visible for a year, flickering ominously with
different colors, and people were truly perplexed about its meaning and nature. Kepler was expecting to see a comet and instead
he got a new star; thus, it was natural for him to suspect that the
conjunction had caused the bright star to burst forth. He calculated that the conjunction of the three planets occurred every 805
years and that they must have gathered together in 6 B.C., close to
the time of Jesus' birth.[9] Although Kepler mused about the conjunction in 6 B.C. and the appearance of a new star, his mysticism
led him to believe instead that the Star of Bethlehem had been a
miracle, not a comet or a new star. He believed that it had been a
special star, one that the Magi could never have foreseen.

This star was not of the ordinary run of comets or new stars, but a special miracle moved in the lower layer of the atmosphere. (*Kepleri opera omnia* 4.346) [10]

Given this evidence, it is puzzling that Kepler's work on the Star of Bethlehem has been used to advocate a supernova as the sign of the birth of Jesus because Kepler himself discounted the theory that a new star was the Magi's star.

In addition, Roman and Greek astrologers were generally not interested in new stars; however, there is one historical account from Roman times about a new star, a star that played an important role in the history of astronomy. Pliny the Elder reported that Hipparchus (d. 126 B.C.) had observed a new star.

Hipparchus . . . having done more to prove that man is related to the stars and that our souls are a part of heaven, detected a *new star* that came into existence during his lifetime; the movement of this star in its line of radiance [brightness change] led him to wonder whether this was a frequent occurrence. (*Natural History* 2.95)

Chinese records show that a nova, called a "guest star," appeared from June 22 to July 21, 134 B.C., in the constellation Scorpius, and this nova may have been the one seen by Hipparchus.[11] Pliny's failure to give us any astrological interpretation of Hipparchus's nova agrees with the ancient literature that these transient stars were not important to astrologers. In fact, the vast collections of ancient astrological treatises by Ptolemy, Dorotheus, Hephaestion, Firmicus, Valens, and others do not examine these fleeting visitors. The extensive Chinese records do report various guest stars, and one was seen in 5 B.C., close to the time of Jesus' birth, but there is no known Western record of that nova.[12] Of course, it is possible that the record of the nova did not survive the ages, but the surviving astrological material still rules out any regal significance. For example, there are no ancient Western horoscopes claiming the blessings of any new star. Thus, it appears that these short-lived stars, which were closely watched by Chinese stargazers, were ignored by Western astrologers of Roman times.

The nova or supernova theory is an excellent example of how theorists rationalize that the accidental coincidence of an astronomical event with the birth of Jesus must have been astrologically interpreted as a regal portent and that, therefore, the event must have been the Star of Bethlehem. This fallacious line of reasoning is called *post hoc, ergo propter hoc* (after this, therefore on account of it). That is, a timely coincidence of two events can give the illusion that the events are related. In any case, there is no historical basis for claiming that a nova signified the birth of a king; thus, this theory for the Star of Bethlehem must be disregarded.

The Planetary Conjunction

Although the Greeks gave us our word "planet" (planoi = *planos* = wanderer), they more often used aster (star) to describe planets. For example, the planet Venus was often called "the star of Aphrodite"; the planet Jupiter, "the star of Zeus"; and so forth. For this reason many explanations of the Magi's star have focused on finding a planet that fits the biblical account. Moreover, planetary configurations are known to have had an astrological meaning, and some planetary conjunctions did indicate regal births according to ancient beliefs. But some investigators have proposed a conjunction as the explanation for the Star of Bethlehem not because astrological configurations were seen as important by the ancients but primarily because conjunctions were visually exciting events.[13]

Like the new star theory, the most famous of the planetary conjunction theories comes from Kepler's analysis of the Star of Bethlehem; however, as we have seen, Kepler thought that the herald of the birth of Jesus had been a miraculous star. He also believed that this supernatural apparition had been accompanied by a new star produced by a triple conjunction. Moreover, Kepler's faith assured him that the Magi would have recognized such an auspicious portent because "God would have condescended to cater to the credulity of the Chaldeans [astrologers]."[14]

Lacking any evidence for Kepler's miracle star or a spectacular supernova, some researchers have advocated a triple planetary conjunction as an explanation of the Magi's star. Modern calculations verify Kepler's finding that there was indeed a triple conjunction early in 6 B.C. Furthermore, a triple conjunction would surely have had astrological significance, and even Kepler remarked on the regal astrological portents offered by a great conjunction of planets.

> The Magi were of Chaldea, where astrology was born, of which this is a dictum: Great conjunctions of planets in cardinal points, especially in equinoctial points of Aries and Libra, signify a universal change of affairs; and a cometary star appearing at the same time tells of the rise of a king. (*Kepleri opera omnia* 4.347)

It is odd that Kepler mentioned Aries: the triple conjunctions that he discusses, in A.D. 1604 and 6 B.C., happened in Sagittarius and Pisces, respectively, not in Aries. In making these remarks Kepler was most likely drawing attention to the power of great conjunctions in important signs such as Aries (a cardinal sign holding the vernal equinox). In spite of his mention of Aries, he did not examine whether there was a possible connection between that zodiacal sign and the Star of Bethlehem. Instead he believed that the miracle star and conjunction were associated with Pisces, which is adjacent to Aries. He also believed that a comet accompanying the conjunction presaged an upheaval by a new king, not the birth of a king. Thus, he saw the conjunction as setting the basis for a new ruler, an idea that must be examined more closely.

Although a triple conjunction seems to be a reasonable explanation for the Star of Bethlehem because it was rare and visually impressive, two issues must be resolved. Did astrologers of antiquity think that this triple conjunction signified a regal birth? And how was the conjunction related to the birth of a king in Judea?

The great astrologer Claudius Ptolemy does not explain (ca. A.D. 150) whether the triple planetary conjunction was as auspicious as Kepler maintains. (Kepler was most likely expressing a Renaissance or personal viewpoint.) But another important astrologer, Vettius Valens, gives the conjunction good, even regal, potential (ca. A.D. 175).

Saturn, Jupiter, and Mars—bring together benefits and make the
native a celebrity, exercising governmental functions, leaders, pro-
tectors, the heads of people and countries; they command and are
obeyed; they are not able to truly avoid their esteemed life, but
are swept up in an adversarial storm of accusations, violence and
living in terror; one must discern these events according to the positions of the houses and the characteristic power of the signs.
(*Anthology* 1.20)[15]

Valens explains that the influence of the triple conjunction depends on the house (the sector in the sky) and signs in which the
conjunction is found. And Firmicus confirms Valens's reservation
by analyzing a special case of a triple conjunction of Saturn, Jupiter, and Mars. His analysis (ca. A.D. 334) shows how a seemingly
good planetary aspect can become a bad portent.

If the Moon is on the Medium Caelum [the Midheaven, or highest
sign in the zodiac], and Saturn, Mars, and Jupiter are together on
the Imum Caelum [the anti-Midheaven, or lowest sign in the zo-
diac], the native will be dragged from his own country and sold into
slavery. (*Mathesis* 6.31.19)

Valens's extensive analysis of triple conjunctions notes only one
group that can reliably produce regal conditions: the Sun, the
Moon, and Jupiter.[16] Thus, it appears that Kepler's triple conjunction of Mars, Jupiter, and Saturn was not an irrefutable portent of
a regal birth, although it does seem to have some potential. Nevertheless, would Kepler's conjunction have pointed to Judea according to ancient practices?

As we have seen, Kepler believed that the triple conjunction in
6 B.C. was associated with Pisces, the Fishes; and calculations confirm this as the correct location. Mars joined Jupiter and Saturn in
Pisces on approximately January 15. On February 5 Jupiter left
Pisces and moved into Aries.[17] Kepler believed that Pisces represented Judea. This idea came primarily from the work of Rabbi
Isaac Abarbanel (A.D. 1437–1508), a Sephardic Jew who, rather
than be forced to convert to Christianity, fled Spain in 1492.[18]
Abarbanel wrote a midrashic interpretation about the prophecy

of the Messiah, and in *The Wells of Salvation* (ca. 1497) he developed his own ideas about interpreting celestial portents that revealed the advent of the Messiah. He claimed that a conjunction of Jupiter and Saturn in Pisces had preceded the birth of Moses by three years. And drawing upon other conjunctions that coincided with the appearance of great Jewish leaders, Abarbanel concluded that a conjunction of Jupiter and Saturn in Pisces had signaled the birth of the Messiah in Judea.[19]

Abarbanel's ideas are problematic for several reasons. First, less is known about the birthday of Moses than that of Jesus. And even if Moses' birth year were known, the delay of three years between the conjunction and his birth would be incongruous with standard astrological principles. Astrologers claimed that the hour of birth, not the year, determined the infant's destiny.[20]

Second, conjunctions between Jupiter and Saturn were not regarded as distinctively regal portents by astrologers of Herod's time. Abarbanel did not understand that to signify a regal birth conjunctions must be accompanied by additional astrological conditions. Thus, it is apparent that Abarbanel drew his conclusions from his personal beliefs, not from primary astrological sources of Roman times.[21]

There is a third and important problem with Abarbanel's speculations. According to Claudius Ptolemy, Pisces was not related to Judea as Abarbanel claimed. The Fishes symbolized Libya (North Africa) and the area from Lydia to Cilicia in southern and southwestern Turkey. No primary sources from Roman times confirm Abarbanel's ideas about Pisces. The notion that Pisces held the Star of Bethlehem received much support among Renaissance Christians because they realized that the vernal equinox had moved from the constellation of Aries into Pisces with the dawning of Christianity. And many people saw a connection between the Fishes and the Christian fish symbol (*ichthys,* the Greek word for "fish," formed an acronym for "Jesus Christ, God's Son, Savior"). Thus, theories that rely on Pisces as the sign of the Jews are erroneously based upon a midrashic interpretation from the Renaissance, an interpretation supported by Christian symbolism.[22] None of these notions accurately reflect the practices and beliefs of the time of the Magi.

Similarly, Kepler's views and conclusions about the Star of
Bethlehem were, as we have seen, a product of his time. That is,
he was interested in the triple conjunction of A.D. 1604 for its
potential to produce comets and for its astrological significance
according to the accepted conventions of the Renaissance. For
example, he referred to the conjunction as the "fiery trigon" (triangle of fire) because the planets gathered in the *trine* (zodiacal
triangle) made by Aries, Leo, Sagittarius—a trine that was associated with the element fire. The three other trines were associated
with earth, air, and water. However, the connections between the
four elements and the trines were not made until the sixth century A.D. and were thus unknown during Herod's time.[23]

Another type of triple conjunction has been suggested as an
explanation for the Star of Bethlehem.[24] This modern explanation
proposes that three individual conjunctions between Jupiter and
Saturn formed the basis for a regal star. The conjunctions took
place over the course of 7 B.C., which is within the anticipated
time frame for Jesus' birth. This explanation is based upon Abarbanel's ideas and holds that according to Babylonian traditions
Jupiter was the planet related to kings and Saturn was the protec-
tor of the Jews. A possible connection between Jews and Saturn
is intriguing in view of the fact that the Roman historian Tacitus
also claimed that Jews were controlled by Saturn because they
worshipped on Saturday (Saturn's day). However, Tacitus's notion that the Sabbath was held holy because of Saturn has no basis
in Judaism. Moreover, there is no convincing evidence that ancient Babylonian traditions were practiced by astrologers during
Jesus' time.[25]

This newer theory suggests that the three conjunctions of Jupiter and Saturn in the sign of Pisces form the basis for the Magi's
star because the planets made a close pass, on the order of one
degree, which is twice the diameter of a full moon. From an astrological point of view these conjunctions could be important, but
it is the astronomical perspective, namely, visible massing, that
the theory emphasizes. The theory argues that the conjunctions
were important because they were visible in the east after sunset.[26] And for those who believe that the biblical account implies
more than one appearance of the Magi's star, the succession of

three conjunctions offers an explanation. However, this theory about the three conjunctions is problematic because the rising of planets at sunset is not what "in the east" meant to astrologers of antiquity. Furthermore, a major pitfall is that these conjunctions happened in Pisces, which was not the zodiacal sign of the Jews.[27] Thus, this triple conjunction is an astronomical theory that does not agree with the principles and practices of ancient astrology.

PERCEPTIONS OF PORTENTS

The problem with all the theories presented so far—comets, supernovas, and planetary conjunctions—is that they are not supported by the historical evidence about ancient perceptions. These explanations of the Magi's star invariably give visual impact the greatest importance, whereas the ancient perceptions are placed secondary and even incorrectly interpreted. Furthermore, little attention is given to what diviners at the time actually would have recognized as a celestial portent of a regal birth. As we have seen, comets were feared; new stars were ignored, or at least not discussed; and planetary conjunctions have been misinterpreted.

An intriguing celestial event occurring in the right time frame is not necessarily a solution unless contemporaneous sources can substantiate it as being a regal portent. Moreover, too much credence is given to the musings of Abarbanel and Kepler during the Renaissance, while the vast astrological record of Ptolemy, Dorotheus, Valens, Firmicus, Hephaestion, and others from Roman times is ignored.

While many scholarly explanations of the Star of Bethlehem have taken astronomical events and wrapped them within an astrological justification, others have applied mythology, often from different cultures. When explanations mix mythologies and beliefs of different cultures, they can produce ideas that are intriguing but nevertheless without a historical basis. For example, one theory noted that because of the mythological connection between the goddess Aphrodite (Venus) and the god Adonis and because of the similarities between the myth of Adonis and the life of Jesus, the Star of Bethlehem may have been the planet Ve-

nus.[28] But this theory is not supported by any of the astrological
works from Roman times.

And sometimes even astrology is mythicized in such a way that
the zodiacal signs are given incorrect meanings and used to im-
part significance to a planetary conjunction that happened in a
certain sign and time frame. For example the zodiacal sign Virgo,
the Virgin, was said to signify the Virgin Mary, Mother of Jesus.
Even Aries, the Ram, was suggested to symbolize Jesus as the
Lamb of God (Agnus Dei), or Jesus as the Good Shepherd.[29] Astrologers of Herod's time would never have made such connections, and in any case such Christian concepts and metaphors
would have been introduced only many years later. There is nothing in the principles and practices of astrology of Roman times
that supports those clever reinterpretations.

It cannot be said often enough that the answer to the puzzle
lies in the kind of astrology that dominated the Roman world at
the time of the birth of Jesus. The search for historical truth must
not ignore the overwhelming evidence that the Magi practiced a
highly Hellenized form of astrology that was not Babylonian and
definitely not Renaissance European.[30] Moreover, personal beliefs
about the validity of astrology must not distort the fact that people
of ancient times, for the most part, respected and admired astrologers, who earnestly pursued knowledge of how celestial events
and bodies affected people. The correct approach to solving the
mystery of the Star of Bethlehem is, therefore, to identify those
astrological conditions that people of ancient times would have
believed to point to a birth of a king in Judea.

But before turning to computers to search for answers, we
must open our minds to what the ancient record of Greek astrology tells us. Our objective is to find a celestial event that unquestionably would have been a portent of a regal birth according to
astrologers of Herod's time—something that would have left the
Magi breathless—something that would have pointed to the Messiah in Judea.

The Dawning of the
First Millennium

FATE'S ACOLYTES

The answers to our questions about the Star of Bethlehem lie in the philosophical and religious practices of the people who interpreted celestial events as portents. Their beliefs about the sky were markedly different from ours and show that the answer can be found not in astronomy but in a form of astrology that partnered closely with powerful beliefs about humankind's relationship with the universe. Putting aside our personal feelings about ancient pagan beliefs, we must strive to understand the perception of celestial portents during Roman times.

To decipher the riddle of the star, we must begin by examining the Magi. Their name comes from the Greek word *magoi,* the plural of *magos,* and is the root of our English word "magician." The title was originally given to a caste of Zoroastrian priests who, in a struggle to take over the Persian throne, were beaten and slaughtered by King Darius (521–519 B.C.). Over time the title lost its connection to Zoroastrianism and was given to seers who predicted the future from omens and dreams. Well before the time of Jesus, the Greek historian Herodotus wrote that a magos was "one of the priests and wise men in Persia" who interpreted dreams. These learned sages were truly esteemed advisors; they served kings and were even believed to have exalted rank. The Christian writer Tertullian said (ca. A.D. 210), "the East considers magi almost as kings," a remark that has made many people believe that the Magi of the Bible were royalty; thus, they became mythicized as the Three Kings.[1]

In ancient times, magi had several important specialties, such as healing the sick and explaining physical phenomena (natural science), but it was their ability to interpret omens that received special acclaim. The magi of the Near East were particularly well known for practicing astrology, which is another form of interpreting natural phenomena as portents or signs of the future. Thus, some biblical scholars translate *magi* as "astrologers," and three of the seven major English translations of Matthew note that the Magi were astrologers or people who practiced astrology.[2] The fact that the Magi of the Bible found a celestial omen that heralded the birth of a king indicates that they were indeed talking about astrology. Therefore, we should consider them to be astrologers.

The principles and practices of interpreting celestial omens and the important role that seers such as the Magi played in the Greco-Roman world of the first century B.C. are revealed in the philosophical and religious ideas of the times. Those beliefs help us to understand the kind of astrology that was practiced at the time of Jesus' birth. Judaism was a small island in a vast Hellenistic sea in which Fate was believed to control everything. The Roman poet and astrologer Manilius gave the famous dictum (ca. A.D. 15) that guided many people:

> Fate rules the world, all things stand fixed by its immutable laws, and the long ages are assigned a predestined course of events. (*Astronomica* 4.14–15)

The conquests of Alexander the Great during the fourth century B.C. accelerated the dissemination of Hellenistic ideas and concepts throughout the Near East, from Egypt to Babylonia. One philosophy in particular, Stoicism, became dominant and was practiced to one extent or another by many people, especially by Romans.

The Stoics taught that the universe was a living being, spherical in shape as manifested by the sky, and that it controlled all events by a divine fire that was "reason." Each person's soul was a small flame from the cosmic fire, which was a divine force, namely, Fate.[3] Thus, Stoics lived "according to nature," which meant that they accepted whatever events Fate sent. The Romans called divine Fate Fortuna, and the Greeks called her Tyche. And temples and monuments in adoration of the goddess of fate were erected throughout the Greco-Roman world—especially in the Near East.

An important idea of those times, and one not associated with Stoicism exclusively, was that divination, the reading of omens, could reveal the will of all-powerful Fate.[4] It was believed that a person could either attempt to ascertain the intentions of Fate (and perhaps even take precautionary steps in some instances) or be oblivious to her signs and suffer the consequences. That idea is strongly conveyed by the popular expression

Fate guides the willing, drags along him who resists. (Seneca *Epistles* 107.10)

Of course, not everyone unflinchingly accepted his fate. We know that people did respond to omens and did take action that they thought would prevent predicted events or at least mitigate the severity of bad portents. And some people used divination to ascertain whether they had divine approval for various courses of action. Furthermore, the reading of omens was not practiced by Stoics exclusively: the majority of the ancient world looked for signs that foretold the future. Such ideas about omens were held by many religions of ancient times (some people even interpreted omens as signs from God). Thus, the Magi and other diviners provided a much needed service by interpreting signs, which gave people guidance and understanding of the events in their lives.

Astrology was a natural response to the quest to understand omens, and more important, it partnered closely with Stoicism, which professed that the universe controlled life. Because astrology wonderfully fit Hellenistic ideas about humankind's relationship to the universe, the Greeks focused much of their intellectual energy on developing and organizing conventions for interpreting the movements of stars and planets.

Although astrology became very popular in the Near East, astrologers were not readily found in Herod's kingdom: Jews did not practice astrology, especially not to the extent that is found in other parts of the Greco-Roman world. Nevertheless, recent archeological findings show that some astrological ideas were adapted by Jews living in Hellenistic communities in the fourth century A.D. and later. In particular, there is scattered evidence of artistic and mythological uses of the zodiac in some synagogues. Also, some documents found in the Dead Sea Scrolls indicate an interest in astrology by a few Jews. Nevertheless, astrology did not assume in traditional Judaism the powerful role that it enjoyed in the countries and cultures surrounding Judea.[5] And that lack of interest is documented:

> [The Jews] do not worry about the cyclic course of the Sun or the Moon. . . .Neither do they practice the astrological predictions of the Chaldeans. (*Sibylline Oracles* 3.218)

Their lack of interest is also evident from the account in Matthew: that Herod, his advisors, and the people of Jerusalem were mystified by the Magi's report of a regal star agrees with the evidence that Jews did not practice astrology. If there were astrologers with important information for the king of Judea, they would have most likely come from another country, one where astrology was actively practiced. Thus, the Magi most likely came "from the East," as Matthew reports, and not from Judea.

Although astrological concepts about predicting a person's fate or destiny were strange to Jews, news of a celestial portent pointing to a new king would nevertheless have attracted much attention among people hoping for the coming of the Messiah. The next question is what kind of celestial portent led the Magi to believe that a king of the Jews had been born. What kind of astrology did they practice?

W I S E   M E N   F R O M   T H E   E A S T

In the account in Matthew, there is little information about the Magi. Contrary to popular opinion, there are no records about how many Magi there were, and their names are not given. In addition no one knows from what country the Magi came; no indication other than "from the East" is given in the biblical account. During Herod's reign, there were many cities and countries to the east of Jerusalem where magi, that is, astrologers, were extremely active. (The eastern origin of astrology was usually associated with Chaldea in Babylonia; therefore, astrologers were also called Chaldeans.)[6] The most likely candidates are Mesopotamia and Babylonia because Western astrology began and thrived there (see map 2). Most researchers believe that the Magi came from this region, which was under the control of the Parthians during the time of King Herod. Early Church leaders such as Clement of Alexandria (ca. A.D. 150–215) and St. Cyril of Alexandria (bishop A.D. 412–44) also favored Parthia, which lay to the east of Herod's kingdom.[7]

In figure 4 a Roman gravestone of the third or fourth century from a catacomb depicts a deceased woman named Severa with the inscription "Severa in Deo vivas" (Severa, mayst thou live in God). To the right of the inscription are the Magi offering gifts to the infant Jesus seated on the lap of his mother, Mary. The star sits above Mary and the infant, and behind them stands either the prophet Balaam pointing to the star or a personification of the Holy Spirit. In any case, this early tablet illustrates well the eastern origins of the Wise Men: they are wearing Phrygian caps, which were related exclusively to eastern people.[8] This floppy hat evolved over the centuries into the pointy, star-studded sorcerer's cap that medieval astrologers are often portrayed as wearing.

Figure 4. The Magi are shown in eastern attire in the Epitaph of Severa, 3d–4th century A.D. *Raffaele Garrucci,* Storia della arte cristiana nei primi otto secoli della chiesa *(Prato: G. Guasti, 1881), vol. 6, plate 485.*

Knowing the Magi's nationality is not as important as understanding the astrological principles and practices that they most likely professed. Moreover, it is misleading to presume that the Magi were Babylonian astrologers. The Magi may have been from the region of Babylonia, but they did not practice archaic Babylonian astrology: that form of astrology had passed into history after the conquests of Alexander the Great.[9] Under the Seleucids, who ruled Babylonia after Alexander's death in 323 B.C., astrology had been transformed into a system that was unmistakably Greek in its theories and philosophy. Thus, the image of the Magi as Babylonian astrologers sitting on top of a ziggurat or observation tower watching for the appearance of a regal stellar omen is historically wrong for the time of Herod.

Babylonian astrology survived for two millennia, and it evolved greatly over that time. Records from ca. 1000 B.C. and earlier show that astrologers extensively observed the skies for the appearance of omens that consisted chiefly of atmospheric phenomena such as meteors and unusually colored sunsets. Astrologers were stationed across the Babylonian Empire watching for celestial portents from observation towers, and daily reports were sent to the king. Because the astrological interpretations were concerned with the fate of the country or its government, namely, the king, Babylonian astrology is called *judicial astrology.* Here is an example of a forecast from Babylonian astrologers about Amurru, a country that lay immediately to the west of Babylonia:

If a meteor train [shower] appears in the north: in the sixth year
the king of the whole world will attack and destroy the king of
Amurru.[10]

Around 750 B.C. records of eclipses and discussions of some
constellations that formed the zodiac made their appearance on
Babylonian cuneiform clay tablets.[11] After ca. 650 B.C. extensive
almanacs listed planetary movements over the course of the year.
In 410 B.C. the earliest known natal astrological chart, nowadays
called a horoscope, was recorded, which describes the positions
of the stars and planets at the moment of a person's birth. The
records from that time also show that the twelve constellations
through which the planets passed became standardized into equal
sectors, called signs.[12]

Although the Greeks and Babylonians had exchanged ideas
and knowledge for centuries, the conquests of Alexander the
Great in 330 B.C. greatly accelerated communication between
these great cultures, and this communication produced major
strides in understanding the workings of the sky.[13] The Babylonians had amassed centuries of celestial observations that they
used to develop mathematical techniques for predicting planetary
motions. But not all of the Babylonian work focused on astrology: their interest was largely in astronomy. For example, the
Babylonians needed accurate predictions of the Moon's phases to
determine the number of days in the lunar months of their calendar, which was needed to establish religious holidays and civic
events. There is evidence that the Babylonians were applying
mathematical analysis to observational data in about 500 B.C.; and
extensive ephemerides, calculated tables of planetary and lunar
positions, date from 205 to 30 B.C. The Babylonian methods for
calculating the risings and settings of the Moon, in particular,
were used by Roman and Greek scientists in the second century A.D.—a fine tribute to Babylonian ingenuity.[14]

The Greeks made important contributions in other areas: they
developed trigonometry and techniques to accelerate the calculation of planetary positions, and Greek celestial globes and charts
improved their ability to locate the stars and planets. In trying to

improve the calculations, the Greeks introduced the concept of modeling the planetary movements as circular motion, which eventually replaced the simplistic, straight-line (zigzag) methods of the Babylonians. More important, this concept about circular motions of the planets laid the foundation for the development of scientific theories about how the solar system worked. Nevertheless, while all of these advances were important for astronomy, astrology benefited even more because astrological charts could now be calculated; that is, all that an astrologer needed to construct a person's horoscope was the time, date, and place of birth.[15]

This background information about the historical development of astronomy and astrology raises an important point concerning the Magi: although they came from the East, they practiced not archaic Babylonian astrology but a newer Hellenistic astrology. The period from 300 B.C. to approximately 150 B.C. is considered to be the formative period for the development of astrology in Western civilization. During that time the Babylonian theories of astrology were greatly revamped and expanded by Hellenistic concepts and philosophy, and Aristotelian science was applied to explain the nature of the astrological influences and their effects. For these reasons scholars call this form of astral divination Greek astrology even though it had some of its origins in Babylonia. To understand the Magi's star, we must realize that the principles and practices of ancient astrology became Hellenistic throughout the Near East and Roman world before the birth of Jesus. Now it becomes apparent why Jews did not embrace astrology: it was a Hellenistic belief that lay apart from traditional Jewish mores and views about the world and God.[16]

Another important fact is that at the time of the birth of Jesus astrologers were calculating the positions of celestial bodies rather than constantly observing the skies to determine their positions, as their Babylonian predecessors had.[17] Observations were still needed to correct mathematical calculations and to verify the occurrence of certain events, like eclipses, that could not be predicted by simplistic techniques.[18] Astrologers, however, did rely upon their mathematics to cast horoscopes. If astrologers watched the sky during a birth, they did so to gauge accurately the time that fixes the orientation of the sky. The part of the zodiac that was rising on the eastern horizon and that determines the time was called the *Horoscopus,* which is the source for our word "horoscope."

The evidence from the preserved record is strong that astrologers of the first century B.C. calculated horoscopes even many years after a birth. One of the most popularized accounts of this practice concerns the calculation of the horoscope of Augustus Caesar, the first Roman emperor. The Roman biographer Suetonius records that in 44 B.C., Augustus (who was eighteen years old at the time) and his friend Marcus Agrippa were in Apollonia, Greece, waiting to join Julius Caesar in the invasion of the Parthian kingdom.[19]

> While in retirement at Apollonia, Augustus mounted with Agrippa to the studio of the astrologer Theogenes. Agrippa was the first to try his fortune, and when a great and almost incredible career was predicted for him, Augustus persisted in concealing the time of his birth and in refusing to disclose it, through diffidence and fear that he might be found to be less eminent. When he at last gave it unwillingly and hesitatingly, and only after many urgent requests, Theogenes sprang up and threw himself at his feet. From that time on Augustus had such faith in his destiny, that he made his horoscope public and issued a silver coin stamped with the sign of the constellation Capricorn, under which he was born [see figure 5]. (*Augustus* 94)

Astrologers also calculated horoscopes using birth and death dates to determine ways to predict life expectancy.[20]

As we have seen, even though there was no correct physical model for the planetary system, the astronomers (astrologers) developed the mathematical means to predict the positions of the planets. Their clever procedure was to match a set of observations made over a long period of time with a mathematical expression. Once they had an expression that gave a reasonable fit for a set of observations, the astronomers used this to predict the skies for

Figure 5. A silver cistophorus (27–26 B.C.) shows Augustus's sign, Capricorn, holding a cornuco- pia signifying the great bounty bestowed upon Augustus by his horoscope. Author's collection, RIC-1 493

other dates. This technique gave Babylonian astronomers a good mathematical description, in particular, of the Moon's extremely complicated motion.[21]

Most calculations of planetary positions were poor by today's standards but were sufficient to satisfy basic astrological require- ments.[22] While the Sun's position and the lunar phases were pre- dicted reasonably well, a planet with a difficult (highly eccentric) orbit, such as Mercury or Mars, could wind up in the wrong zodia- cal sign. Of course, the quest to improve predictions of plane- tary positions is a major theme of the development of Western science.

Having this mathematical capability the astrologers became important seers. And unlike most diviners, who predicted only a day's events by reading the auspices as they happened, astrolo- gers claimed they could forecast the omens and, thus, know in advance the course of events for a complete lifetime. Future days and times were selected by astrologers for celestial conditions that ensured a successful outcome of a venture or endeavor. Moreover, the astrologers' command of mathematical calculations that reasonably predicted the mysterious motions of the planets in the sky bolstered their credibility. Hence, astrologers were also called *mathematici* (mathematicians). Whatever their name, they were considered to be savants; and as astrologers, the Magi of the Bible were indeed respected "wise men" whose services were needed by kings. Most important, the Magi did not watch the skies for a regal star as Babylonian astrologers had done in earlier times. Running through calculations, the Magi saw something auspicious according to the conventions of Greek astrology, something that revealed a newborn king of Judea.

The Sign of the Jews

Within the system of Greek astrology were established rules and conventions that defined regal births and even specified which country was blessed by such a birth. These ideas were derived from studies of how the Sun, Moon, and planets moved among the stars. In ancient times astronomers theorized that the sky was the inside of an immense sphere that held the stars and planets. The earth was thought to be the center of the great celestial sphere whose slow rotation produced the daily rising and setting of the celestial bodies. Those stargazers also noted that the paths of the Sun, Moon, and planets were restricted to a narrow band that wrapped around the sky. This band was called the zodiac, a term derived from the Greek ($\zeta\omega\delta\iota\alpha\kappa\grave{o}\varsigma$ $\kappa\acute{u}\kappa\lambda o\varsigma$ = *zodiakos kuklos* = circle of animals). The stars along the zodiac were grouped into mythical figures, namely, constellations, and were called *zoidions* (signs), which means a picture of a living creature.[23] The signs represent humans, animals, and mythical creatures, which came primarily from ancient Babylonia but were given Greek mythological interpretations.

The middle of the band of the zodiac is marked by the path of the Sun, which is called the *ecliptic* because eclipses occur when the Moon is on this circle and in line with the Sun. Figure 6

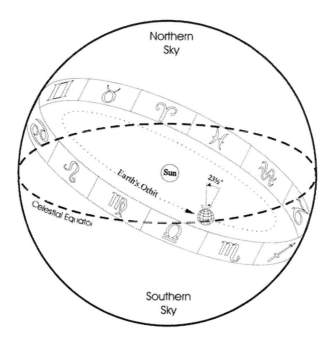

Figure 6. The zodiac is inclined on the celestial sphere because the earth's axis of rotation is tilted with respect to its orbit around the Sun.

shows the sky as a sphere surrounding the earth's orbit. (If the figure were drawn to scale, the sphere would be infinitely larger than the earth's orbit.) On the so-called celestial sphere the zodiac is tilted such that half extends into the northern sky, the other half into the southern sky. The tilt is due to the earth's axis of rotation, which is not perpendicular to the plane of its orbit around the Sun but instead inclined 23½° with respect to that plane. The motion of the Sun through the zodiac is, of course, due to the revolution of the earth around the Sun, and the daily motion of the sky is due to the rotation of the earth on its axis.

People on the earth see the Sun against the background signs in the zodiac, and in the configuration shown in figure 6 the Sun would appear in the sign of Taurus ♉, the Bull.[24] As the Sun moves out of Taurus and into Gemini ♊, the Twins, it climbs higher in the northern sky. The appearance of the Sun in the northernmost part of the zodiac, Cancer ♋, the Crab, marks the start of summer. The Sun then turns south, passing through Leo ♌, the Lion, and Virgo ♍, the Virgin. The Sun then crosses the celestial equator (the boundary between the southern and northern hemispheres of the sky) and enters Libra ♎, the Scales, which marks the beginning of autumn. The Sun proceeds farther south through Scorpio ♏, the Scorpion, and Sagittarius ♐, the Archer.[25] Winter commences when the Sun reaches the southernmost point, in Capricorn ♑, the Sea-Goat. Next the Sun moves though Aquarius ♒, the Water-Bearer, Pisces ♓, the Fishes, and into Aries ♈, the Ram, where it again crosses the celestial equator. This crossing marks the vernal equinox, or the first day of spring, when the Sun passes into the northern sky.

Figure 6 illustrates the orientation of the zodiac two millennia ago. However, the location of the seasons has shifted because of the precession of the equinoxes—which results from the very slow gyration of the earth's axis with respect to the plane of its orbit, a motion similar to the wobbling of a spinning toy top. Figure 7 shows how the celestial equator moves through the constellations along the zodiac as the earth's axis of rotation precesses. Today, spring occurs on the border between the constellations of Pisces and Aquarius, the so-called dawning of the Age of Aquarius (see Appendix A).

The astrologers observed that as the Sun moved through the zodiac during the year, its life-giving radiance changed according to the angle at which the sun's rays struck the earth. Thus, they believed that the Sun's physical effects (heat and light) were evidence for a power emanating from all celestial objects. That this cosmic power went beyond physical heating and luminous stimulation and magically controlled the activities of life and human affairs is a fundamentally important belief to practitioners of astrology.

This belief, along with the philosophy of astrology, is developed in the *Tetrabiblos* (In Four Books), which was written in ca. A.D. 150 by Claudius Ptolemy of Alexandria, Egypt. His authoritative works on astronomy and geography stand as major historical contributions to these fields, and his *Tetrabiblos* is often cited as

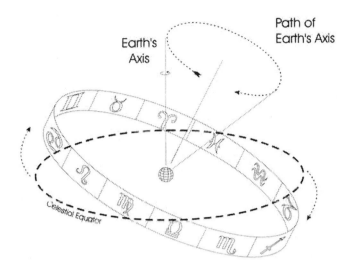

Figure 7. Precession of the earth's axis causes the celestial equator to move through the zodiac and in turn moves the seasons into different constellations over the millennia.

the bible of astrology because other astrologers of antiquity referred to it as an essential reference. Scholars believe that parts of Ptolemy's *Tetrabiblos* were taken from first century B.C sources, which makes it an ideal primary reference concerning astrology during the time of King Herod.[26]

In the *Tetrabiblos,* Ptolemy is often concerned with theoretical analyses of the workings of astrological effects: he frequently explains the rationale and justification behind astrological concepts, and his explanations reflect his analytical, scientific reasoning. In the first book Ptolemy explains the Hellenistic philosophy and theory of astrology—that is, that nature could be explained by pure logical rationalization. Ptolemy first focuses on the obvious seasonal and climatic influences of the Sun and the tidal influences of the Moon:

> For the sun . . . is always in some way affecting everything on the earth, not only by the changes that accompany the seasons of the year to bring about the generation of animals, the productiveness

of planets, the flowing of waters, and the changes of bodies, . . .
The moon, too . . . bestows her effluence. . . .the seas turn their
own tides with her rising and setting, and plants and animals in
whole or in some part wax and wane with her. (*Tetrabiblos* 1.2)

In the second book there is a lengthy and detailed explanation
of how countries were assigned to the control of various signs of
the zodiac. Ptolemy claims that a country's climate (determined
by the Sun) governs the nature of the inhabitants:

The demarcation of national characteristics is established in part by
entire parallels [latitude] and angles [longitude] through their positions relative to the ecliptic and the sun. . . .the people who live
under the more southern parallels . . . have the sun over their heads
and are burned by it, have black skins and thick, woolly hair, . . .
[are] sanguine of nature, . . . savage because their homes are continually oppressed by heat. . . .Those who live under the more
northern parallels are white in complexion, straight haired, . . . and
somewhat cold by nature . . . savage in their habits because their
dwellings are continually cold. (*Tetrabiblos* 2.2)

The countries that Ptolemy goes on to list are those of the first
century B.C. rather than those of his own time, the second century
A.D. The nature of his list indicates that he was indeed drawing his
information from older astrological sources.[27] The first sign, Aries,
the Ram, is related to a number of countries, but he mentions several in the Near East:

The inhabitants of Coele Syria, Idumea, and Judea are more closely
familiar to Aries and Mars, and therefore these peoples are in general bold, godless, and scheming. (*Tetrabiblos* 2.3)[28]

Then, Ptolemy summarizes the assigned countries in a list:

Aries: . . . Coele Syria, Palestine, Idumea, Judea. (*Tetrabiblos* 2.3)

Coele Syria ("Coele," pronounced "koyl-ay," means "valley") was separate from Syria, and although its boundaries varied over the centuries, they generally ranged from the southern part of modern-day Syria, the area of Damascus, westward toward the eastern border of Lebanon (Phoenicia) to slightly south of the present-day Golan Heights (west of Trachonitis), which means that the southern part of Coele Syria was within Herod's kingdom (see map 1).[29] Thus, all of the lands listed by Ptolemy were parts of Herod's realm, which is powerful evidence that astrologers would have monitored Aries, the Ram, for portents about his kingdom and the fate of Jews.

Vettius Valens of Antioch recorded that Aries controlled "Coele Syria and its adjacent lands."[30] Valens was a late contemporary of Ptolemy, and his *Anthology* (ca. A.D. 175) is another important primary source on Greek astrology. His assignment of Coele Syria and its neighbors to the sign of Aries is strong evidence in support of Ptolemy's claim in the *Tetrabiblos* that Herod's kingdom was represented by Aries. By the time of Ptolemy and Valens, the Roman map of the world had markedly changed since King Herod's reign: after A.D. 135 Judea no longer existed as a country but was incorporated into a Roman province called Syria Palaestina. Thus, no other astrological sources refer to Judea.

There is another contemporaneous source that gives zodiacal assignments for various parts of the Roman world. The Roman astrologer Marcus Manilius composed a didactic poem, *Astronomica,* about Greek astrology close to the time of Augustus Caesar and Tiberius (ca. A.D. 10–15). His assignment of countries in the Near East ignores Judea completely (as well as Coele Syria and the other lands mentioned by Ptolemy); he placed Syria and northern Egypt under the control of Aries.[31] Perhaps Manilius, essentially a poet, was not interested in details that Ptolemy and Valens cover in depth; thus, on Manilius's scale of interest Judea did not exist, and Aries represented the entire Near East. It is also likely that the incorporation of Judea and Samaria in A.D. 6 into the Roman province of Syria destroyed Judea's individuality in Manilius's eyes.

In any case, the evidence from three major astrological sources, Ptolemy, Valens, and Manilius, points to Aries as representing Herod's kingdom.[32] If there was a portent involving Judea, it appeared in Aries.

THE APPEARANCE OF THE RAM

During most of the period of the Roman Empire, the cities and client kingdoms under Roman dominion issued their own coinage. Latin inscriptions were used on official imperial issues for the provinces and most cities, but in the eastern part of the empire, in Hellenistic cities such as Antioch in Syria, coin inscriptions were in Greek.[33] The portrait of the emperor or a local deity was usually shown on the obverse (heads side), but the reverse varied: various political or religious themes, often of significance to the local inhabitants, appeared. Above all, Roman provincial coinage served as a primary medium for disseminating propaganda supporting the goals of Rome.[34]

Of the many themes that were used on local coinage, celestial and astral symbols often appeared, mostly stars or crescent moons. But there were also zodiacal symbols. And one of the most common of the zodiacal signs that appeared on coinage throughout the Roman Empire was Capricorn, the birth sign of the emperor Augustus Caesar. Capricorn was even used by later emperors because it was linked to Augustus Caesar and represented a divine right to rule the empire. There are also instances when cities and countries identified themselves with a specific sign of the zodiac on their coinage. For example, Antioch used Aries, the Ram, on its coinage for two centuries. Most important, the first time that Antioch used Aries on its coins coincided with the Roman annexation of Judea and Samaria.

When Herod died in 4 B.C., Rome divided his kingdom among some of his sons, but none of them was given the title king. The Romans were sensitive about calling anyone a king of the Jews because the title potentially had a Messianic message. Although the Romans had been confident of Herod's allegiance, his sons would have to prove themselves; thus, Archelaus was made *eth-*

narch and received the largest and most important part of the kingdom, namely, Judea, Samaria, and Idumea. Two of Archelaus's brothers were awarded smaller kingdoms and were called *tetrarchs*. Herod Antipas (r. 4 B.C.–A.D. 39) received Galilee and Perea (east bank of the Jordan), and Philip (r. 4 B.C.–A.D. 34) ruled over lands in southern Syria (parts of Coele Syria). However, Archelaus's control over Judea was tenuous and constantly threatened. He did not have the firm command to rule as his father did.

Rome eventually overthrew Archelaus in A.D. 6 because of complaints about his abusive rule and banished him to Vienna, a small town in Gaul (see map 2).[35] After removing Archelaus, Augustus ordered the governor of Syria, Publius Sulpicius Quirinius, to oversee the annexation.[36] Quirinius traveled from Antioch to Judea, where he auctioned the property of Archelaus and conducted a census for levying taxes on Jews. Augustus converted the ethnarchy of Judea, Samaria, and Idumea into a province under the rule of a prefect who was a Roman knight (*eques*). Returning to Antioch, Quirinius left behind the new prefect, Coponius, to manage the government of Judea. The prefect had been appointed by Augustus and was responsible for the daily administration of the province. But the main power was held by the imperial governor (*legatus,* "legate") in Antioch, who controlled the legions; the prefect had only some auxiliary units for policing the province.

Aries, the Ram, makes its debut on coins that were issued in Antioch close to, if not during, the time of Quirinius's governorship.[37] One side portrays a bust of Zeus (Jupiter), who had appeared on many coins from Antioch in earlier years, but the reverse illustrates for the first time a leaping ram looking backwards at a star. The scene is truly intriguing in light of the biblical account of Luke 2:1–18 that tells us that Jesus was born when Quirinius was governor of Syria and that there were "shepherds abiding in the field, keeping watch over their flock by night."[38] Of course, the Romans of Antioch were not depicting the Nativity of Jesus. The ram on the coin unmistakably symbolizes the zodiacal sign Aries.

The mythological lore about Aries explains why the ram is

portrayed as leaping and looking backwards. According to Greek mythology a golden ram was sent by the goddess Nephele to save her children, Phrixos and Helle, from being murdered. Phrixos and Helle climbed onto the ram, which then leaped with them into the sky; however, Helle looked down, became dizzy, lost her grip, and fell to her death in the water that divides Greece and modern Turkey, which became known as the Hellespont. Phrixos made it to safety and sacrificed the golden ram to Jupiter, who placed its Golden Fleece among the stars, where it became Aries. The Golden Fleece then became part of the story about Jason and the Argonauts. The lore says that Aries is looking backwards for Helle, and another interpretation says that Aries is admiring his Golden Fleece.[39]

Table 3.1 lists the coins of Antioch that first used Aries, the Ram. The earliest coin is undated, and its monogram ANT is considered to be an abbreviation for Antioch (see figure 8). Numismatists estimate from the coin's style and other evidence that it was issued sometime in the period A.D. 5–11.[40] Thus, it is possible

Table 3.1. Antioch's First Coins Bearing Aries

DATE	OBVERSE	REVERSE
ca. A.D. 5–11	Bust of Zeus	Aries with star above; monogram ANT; "Of the people of the Metropolis of Antioch"
A.D. 11–12	Bust of Zeus	Aries with star above; date BM (42 Actian Era); "Of the people of the Metropolis of Antioch"
A.D. 12–13	Bust of Zeus	Aries with star above; date ΓM (43 Actian Era); "In the magistracy of Silanus—of the people of the city of Antioch"
A.D. 13–14	Bust of Zeus	Aries with star above; date ΔM (44 Actian Era); "In the magistracy of Silanus—of the people of the city of Antioch"

SOURCE: *Andrew Burnett, Michel Amandry, and Pere Pau Ripollès,* Roman Provincial Coinage *(London: British Museum, 1992), vol. 1, 4265–6, 4268–9.*

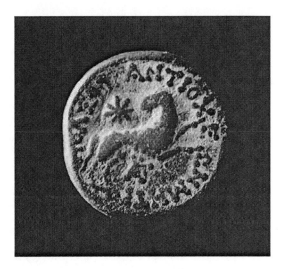

Figure 8. The first bronze coins of Antioch depicting Aries were issued in ca. A.D. 5–11, around the time when Quirinius was governor of Syria. Author's collection, RPC-1 4265

that this coin was minted by Quirinius, who served as Antioch's governor from A.D. 6 to after A.D. 7. On the second coin, the monogram is replaced by the Greek letters BM, which stand for the date: "two and forty" years after the battle of Actium, in which Augustus Caesar defeated Antony and Cleopatra in 31 B.C. This places the coin's minting in A.D. 11–12. The governor who issued this coin is unknown. It could have been Quirinius, an unrecorded governor, or even Q. Caecilius Metellus Creticus Silanus, governor of Syria during the period A.D. 12–17.

The third and fourth coins are almost identical to the first two except that they give the name of the imperial governor of Syria.[41] The Greek inscription translates to "in the magistracy of Silanus — of the people of the city of Antioch." The third coin is dated ΓΜ, "three and forty" years, and the fourth ΔΜ, "four and forty" years (see figure 9); and these were struck in A.D. 12–13 and A.D. 13–14, respectively.

Figure 9. The last coin by the governor Silanus showing Aries was issued in A.D. *13–14. Author's collection, RPC-1 4269*

The historical record indicates that Quirinius became governor of Syria in A.D. 6, but it is unknown when his governorship ceased.[42] Silanus may have been his immediate successor, or there may have been an unrecorded intermediate governor. It is likely, but not provable, that Quirinius had some involvement with the first coin portraying Aries.

In any case, the reason for the use of Aries is perhaps more discernible and lies, in part, in the history of the region. First, the kingdom of Alexander the Great had split apart upon his death, leaving the Seleucids in control of Syria and Babylonia and the Ptolemies in control of Egypt. These two Hellenistic kingdoms fought over Coele Syria and Judea, which lay between them. So, the Roman annexation was extremely important to Syrian prestige from a historical perspective: it represented the achievement of a long-desired goal. Second, the use of a zodiacal sign—namely, Aries, which symbolized their destiny—illustrates the importance of astrology to these people. Placing Aries on the coins may also have served as a means of expressing nationalistic pride in the annexation. That is, Antioch's Romans now had greater control over the lands ruled by Aries. Thus, it is highly likely that the appearance of Aries on Antioch's coins was related to the annexation of Archelaus's kingdom.

For now, interpreting the coins precisely is less important than realizing that these coins initiated my investigation into the Star of Bethlehem. When I examined the astrological sources for possible interpretations of the coins, I found that these sources indicated that the appearance of a star in Aries, not Pisces, would have signified the birth of a king in Judea. But before searching for what appeared in Aries we must ask when it appeared.

THE ERROR IN THE ERA

As King Herod did, many people have asked when the star that marked the birth of Jesus appeared. Most biblical scholars believe that Jesus was born during Herod's reign, but Herod died in 4 B.C. Thus, Jesus was born not at the beginning of the Christian Era but some years earlier. First, we must examine how the calendar was established, which affects the determination of when Jesus was born.

When Julius Caesar arrived in Egypt in 48 B.C., he found that astronomy there had greatly advanced beyond what Rome knew. Sosigenes, a Greek astronomer from Alexandria, introduced him to many scientific findings, in particular, to a calendar that was simpler and far more accurate than the Roman calendar. The Egyptians had known for many centuries that the year—the time the Sun took to traverse the sky—was 365¼ days long. However, the old Egyptian calendar consisted only of 12 months of 30 days each, to which an extra 5 days of festivals (*epagomenal days*) were added to bring the total to 365 days. This meant that the calendar's year was missing one-quarter of a day and that the calendar eventually moved out of phase with the seasons. Despite the error, the Egyptians used this calendar system for centuries and disregarded what we call a tropical, or seasonal, calendar.

The Egyptian astronomers monitored the slow shift of the seasons by watching for the *heliacal rising* of Sothis, known nowadays as Sirius, the Dog Star. A heliacal rising occurs when a star (or planet) is first seen in the morning skies after having spent several weeks obscured by the Sun's brightness. The time between a star's heliacal risings determines the time it takes for the Sun to make one full circuit among the stars—a *sidereal* year. The heliacal rising of Sothis, the brightest star in the sky, also coincided with the annual flooding of the Nile, which was an important event for the Egyptians because the nutrient-rich waters renewed the farm fields in the flood plain. When the heliacal rising of Sothis fell on the Egyptian new year's day, the first of the month of Thoth, a new Sothic Cycle (1,461 years) began. However, the old Egyptian calendar was eventually modified.

In 332 B.C., Greek scholars followed the army of Alexander the Great to Egypt, and the amalgamation of Greek and Egyptian cultures produced an explosive growth of knowledge and led to great advances in astronomy. In 237 B.C., during the reign of King Ptolemy III, a convocation of priests issued the Decree of Canopus, which was written on a stele in Greek, Egyptian hieroglyphics, and demotic, a stele similar to the famous Rosetta Stone issued forty-three years later by King Ptolemy V. The Decree of Canopus announced the adoption of a new Egyptian calendar that incorporated a leap year.

> And that the seasons of the year may coincide with the constitution of the world, and that it may not happen that some of the popular festivals which ought to be held in the winter come to be celebrated in the summer, owing to the Sun changing one day in the course of four years, . . . from this time onward one day, a festival of the Well-Doing Gods [King Ptolemy and Queen Berenice], shall be added every four years for the five additional days, before the New Year, so that all may know that the error of deficiency which existed formerly in respect of the arrangement of the seasons, and of the year, . . . hath been rectified. (*Decree of Canopus*)[43]

In 46 B.C., nearly three centuries after the Decree of Canopus, Julius Caesar recognized the elegant simplicity and accuracy of the new Egyptian calendar, and he incorporated the leap year concept in the new Roman calendar, which we call the Julian Calendar. In creating his calendar, he added two months as a temporary measure to get the seasons to agree with the customary Roman dates. Most important to our determination of the birthday of Jesus, Caesar set December 25 as the winter solstice, the first day of winter. His calendar started on January 1, 45 B.C., and remained the standard calendar in Western civilization until 1582, when a slightly modified version known as the Gregorian Calendar was instituted.[44]

Although the calendar we use today was essentially in use before the birth of Jesus, the era system, that is, the designations B.C. and A.D., was not implemented until much later. In A.D. 533, Dionysius Exiguus (Dennis the Little), a Christian monk in Rome, proposed that the enumeration of the years should be based upon the birth of Jesus. Dionysius added up the lengths of the reigns of the Roman emperors since the year of Jesus' birth, which was in Dionysius's time celebrated on December 25. He set December 25, 1 B.C., as the birthday and defined the start of the Christian Era one week later, on January 1, A.D. 1.[45] As a result, the year of Herod's death became 4 B.C. Since the historical evidence indicates that Jesus was born during Herod's reign, Dionysius evidently made a mistake in summing the reigns of the Roman emperors. (Even scholars of his time argued over how many years he had skipped.)

Another error or misconception that Dionysius perpetuated was the fixing of December 25 as the birthday of Jesus. The problem Dionysius faced is that the correct date, if it had ever been known, had definitely been forgotten by his time. How the date was set as December 25 is a complicated story. In the early third century Clement of Alexandria (ca. A.D. 150–215) wrote about theories that Jesus had been born in the spring, on April 20–21 or May 20.[46] In the fourth century Christians celebrated January 6 as the Epiphany of Jesus, which became associated with the Adoration of the Magi. However, the early Church was more interested in celebrating other dates in the life of Jesus, such as the Crucifixion, the Resurrection, and his baptism; and the evidence indicates that early in the history of the Church there was no established birthday for Jesus.

The earliest reference to a Nativity celebration on December 25 comes from a Roman document known as the Philocalian Calendar, which dates from A.D. 354 but may include an earlier reference to A.D. 336.[47] Some people erroneously think that the feast day of Jesus' birth was set to coincide with the Saturnalia, a Roman festival held at the close of autumn in honor of Saturn, the god of farming and the harvest. However, the Saturnalia, which fell on December 17 and lasted three to seven days, ended before December 25. Although historians note that the merriment of Christmas Eve and New Year's Eve may in fact be a vestige of this ancient Roman holiday, they note two much more important reasons why Christians would have selected December 25 for Jesus' birthday.

As it happens, pagan Romans celebrated two significant holidays on December 25. The first was one of the most significant holidays and was even grander than the Saturnalia. In A.D. 275 Emperor Aurelian had decreed that December 25 was *Dies Natalis Solis Invicti* (the birthday of the Unconquerable Sun). *Sol Invictus* was called "the heart of the universe," because the Sun played significant roles in nature and because according to the Stoics the Sun was the central divine fire of the universe. The radiance of the Sun regulated life on the earth seasonally as well as daily, and the Roman emperor was often considered as the Sun incarnate—the essential, central source of Roman civilization. Furthermore, the Sun's motion among the stars defined the path of the zodiac, to which the motions of all the planets and the Moon were confined. Thus, the Sun was worshipped for its influence and the emperor shared in this glory.

The second celebration on December 25 was connected to the secret cult of Mithras, which was very popular among the military across the empire. The Roman fascination with astrology eventually led to the incorporation of aspects of astrology into religious beliefs, and Mithras was a celestial god who was evidently seen as responsible for the precession of the equinoxes.

According to the Mithraic art that has survived the ages, Mithras and the Unconquerable Sun ruled over the cosmos together; that is, they oversaw and controlled the motions of the stars and planets.[48]

Many scholars believe that December 25 was chosen as Jesus' birthday because Christians wanted to give a Christian meaning to the popular pagan holiday, and it is readily apparent that they succeeded, perhaps beyond their wildest expectations. Today December 25 is called Christmas in English, a name that comes from medieval times, when the "festival of Christ" was known as *Christes maesse* (Christ's Mass). However, biblical scholars agree that December 25, 1 B.C., the date that Dionysius selected, does not mark the birth of Jesus, and more evidence is needed to get a better estimate for his birth.

THE FIRST CHRISTMAS

The error in establishing the start of the Christian Era, along with the lack of a recorded date for the birth of Jesus, requires that we find some method to determine the date of his birth. There are several events and accounts that we can use to establish a time period. In particular, most biblical scholars believe, according to the account in Matthew, that Jesus was born during the reign of King Herod; therefore, Herod must have died after Jesus was born. However, this benchmark is problematic because the date of Herod's death is itself the subject of controversy among some researchers.

Our information about the death of Herod comes from the historian Flavius Josephus (A.D. 37–ca. 100), who reports in *Antiquities of the Jews* that Herod died after an eclipse of the Moon and was buried before Passover. There was a lunar eclipse on the night of March 12–13, 4 B.C., which was a month before Passover on April 11, 4 B.C. While most historians overwhelmingly favor the year 4 B.C., a few researchers have questioned it because there were other years when an eclipse of the Moon preceded Passover.[49]

Some suggest the eclipse in 1 B.C. as an alternative because this eclipse happened in January, which allows more time for the several events that Josephus cites as occurring between the eclipse and Passover.[50] The eclipse of 1 B.C. is favored by some researchers for another reason as well: the eclipse of that year was more impressive than the one of 4 B.C. The lunar eclipse of 1 B.C. was total and readily visible, whereas the one of 4 B.C. was unimpressive, a partial eclipse in which only two-fifths of the moon was darkened. Furthermore, the eclipse of 4 B.C. was also difficult to see because it occurred between 2 and 4 A.M. in Jerusalem.

Those who discount the eclipse of 4 B.C. for these reasons fail to recognize the literary purpose that the eclipse serves in warning about Herod's impending death.[51] Furthermore, proposing a more impressive eclipse demonstrates a failure to appreciate how people of antiquity perceived omens. An eclipse would have frightened people whether they saw it or not. Josephus noted it not because it was conspicuous but because eclipses of any kind were omens. He was echoing a widely held fear of eclipses, and he applied that fear to the story about Herod, who was soon to undergo a terribly painful death. Interestingly, Hephaestion of Thebes (an astrologer of ca. A.D. 415) recorded that a lunar eclipse warned "that a great tyrant will be depressed and will have his house in disorder."[52] No prognostication could be more accurate about Herod. Therefore, there is no compelling reason to move Herod's death from late March or early April of 4 B.C. If we accept the account in Matthew, this time is the latest possible date for the birth of Jesus.

Some researchers have tried to apply the census reported in Luke 2:1–3 as another benchmark to establish the time of Jesus' birth.

> And it came to pass in those days, that there went out a decree from Caesar Augustus, that all the world should be taxed. (*And* this taxing was first made when Cyrenius was governor of Syria.)
> All went to be taxed, every one into his own city. (Luke 2:1–3)

In reporting the events surrounding the birth of Jesus, the account in Luke does not mention Herod or the Magi. Instead Luke cites Cyrenius (Quirinius), the governor of Syria, as conducting a census under orders from Augustus Caesar. (This is the same Quirinius who may have issued the first coins of Antioch bearing Aries.) The trouble with using the census of Quirinius as a benchmark for establishing the birth of Jesus is that the census was conducted after Archelaus was deposed in A.D. 6, ten years after the generally accepted date for Herod's death. But, as we know, the account in Matthew reports that Jesus was born during Herod's reign.

Researchers who have tried to resolve this discrepancy have proposed that Quirinius may have held another, earlier governorship in Syria, during Herod's reign; however, the historical evidence does not support that idea.[53] Table 3.2 shows when the imperial governors (legates) of Syria held office. The records indicate that Publius Sulpicius Quirinius served as governor of Syria only once, well after Herod died.[54] Some researchers have suggested that Quirinius may have been governor from 4 to 1 B.C., the period now believed to have been held by L. Calpurnius Piso. However, most historians maintain that Quirinius held only one governorship of Syria, which started in A.D. 6 and ended sometime in the period from A.D. 7 to 12.

In addition to falling outside the reign of Herod, the census of Quirinius presents irreconcilable differences with other chronological information given in Luke, information that can be used to

Table 3.2. *Roman Governors of Syria*

RULE	GOVERNOR
23–13 B.C.	M. Agrippa
ca. 12–10 B.C.	M. Titius
9–6 B.C.	S. Sentius Saturninus
6–4 B.C.	P. Quinctilius Varus
4–1 B.C.	L. Calpurnius Piso (?)
1 B.C.–A.D. 4	Gaius Caesar
A.D. 4–ca. 5	L. Volusius Saturninus
A.D. 6 to after 7	P. Sulpicius Quirinius
A.D. 12/13–17	Q. Caecilius Metellus Creticus Silanus

SOURCE: *Bengt E. Thomasson*, Laterculi Praesidum *(Arlöv: University of Gothenburg, 1984), 1:303–5, 2:3; and Ronald Syme*, The Roman Revolution *(London: Oxford University Press, 1963), 397n. 99, 437.*

establish the time period for the birth.[55] Specifically, Luke 1:5 indicates that the annunciation of the conception of John the Baptist was "in the days of King Herod of Judea," and Luke 1:36 puts Jesus' birth fifteen months later. This means that Jesus was born about six months after John the Baptist. If we assume that John was born when Herod died, in April 4 B.C., the latest that Jesus could have been born was later that year, in October. This bit of information serves as a third benchmark for the birth of Jesus.

Another benchmark comes from Luke 3:1, which says that John the Baptist began preaching in the fifteenth year (A.D. 28–29) of the reign of the emperor Tiberius, and from Luke 3:23, which says that Jesus started his ministry after that time, when he was "about thirty years old."[56] If we interpret "about thirty" as anywhere from twenty-seven to thirty-three, then the birth must have fallen between 6 B.C. and A.D. 2 This is consistent with a birth in 4 B.C. or earlier, and many biblical scholars note that it supports a birth well before Quirinius became governor of Syria in A.D. 6.[57]

Even though the preponderance of evidence indicates a birth in 4 B.C. or earlier, some researchers have continued trying to resolve the conflict between Quirinius's census and the reign of King Herod. First, they have argued that the account of Luke must refer to a census made before the census of Quirinius.[58] This argument is based upon a translation of the Greek word *protos* not as "first" but as "prior"; that is, that the enrollment was not first made by Quirinius but made prior to the census of Quirinius. However, none of the seven most popular English translations of the New Testament accepts this alternative interpretation: all translate *protos* as "first."[59] That is, this was the first Roman taxation enrollment of Judea and Samaria, and the historical records indicate that it was indeed the first.

Second, other investigators note that Augustus Caesar conducted three empirewide censuses: in 28 and 8 B.C. and in A.D. 13–14. However, these censuses were intended for counting Roman citizens living in the provinces, not for taxing people. The counting of people and assessing their wealth for the purpose of levying a tax, as was done in Judea by Quirinius, was not a customary Roman practice. Taxation was carried out in different ways across the empire, but usually a fixed, equal tax was assessed on each city, which in turn levied a tax on landowners. Also, farmers tithed the sale of their crops. Moreover, the system usually allowed some people to escape taxation or to pay only a pittance. When Augustus decreed this new tax system, he was introducing a fairer system that called for the orderly registration of everyone. He had first tried this in 27 B.C. in Gaul, where the registration had produced rioting. Quirinius was implementing this new tax enrollment method in Syria when he was ordered to do the same for Judea.[60] The enrollment of A.D. 6–7 in Judea, like that in Gaul, was accompanied by rioting fomented by Jewish religious beliefs that forbade any census. In any event, Augustus's censuses for counting Roman citizens in 28 and 8 B.C. and A.D. 13–14 were not for establishing taxes and would not have required Jews to return to their birthplace as reported in the account of Luke. Quirinius's enrollment was for taxation, which indicates that it was not related to the three censuses of Augustus.

Nevertheless, there is some evidence for possible censuses closer to the birth of Jesus. Writing in A.D. 207–11, the Christian convert Tertullian tells us about censuses taken before Herod died:

> At that time there were censuses that had been taken in Judea under Augustus by Sentius Saturninus, in which they may have enquired about Jesus' ancestry. (*Adversus Marcion* 4.19.10)[61]

To what censuses could he have been referring? Sentius Saturninus was governor of Syria from 9 to 6 B.C., a period that falls during the reign of Herod; however, there is no record of an enrollment for taxing Jews during the reign of Herod (or of Archelaus). And, as we have seen, the census of Augustus Caesar taken in 8 B.C., during Saturninus's governorship, was intended only to count the number of Roman citizens. As Tertullian notes correctly, a person's citizenship would have been examined in Augustus's censuses, but such an examination would not have required non-Romans to travel to their hometowns.[62] Also, it is unlikely that Tertullian confused Saturninus's census with the enrollment mentioned in Luke: Quirinius not Saturninus was cited in Luke. Thus, researchers suspect that the remark by Tertullian suggests that he

knew or believed that Jesus had been born during the governorship of Sentius Saturninus. And he may have presumed that one
of the censuses taken during that time was the census that made
the family of Jesus go to Bethlehem. In any case, Tertullian's remark will be used as another benchmark for the birth of Jesus: 9
to 6 B.C.

Finally, there is the account in Matt. 2:16 of Herod's purported
murdering of two-year-old children. After discovering that the
Magi had failed to identify the recipient of the star's blessing,
Herod ostensibly estimated the age of the child and had those in
that age range killed. However, it is truly curious that children two
years old *and younger* were marked for death: Herod, or more
likely the evangelist, must have failed to understand that the
threat was only from those born on an auspicious day (or group
of days)—not during a two-year interval. And for that reason this
story has been considered to be dubious.[63] Other researchers discount the Slaughter of the Innocents as a myth because Josephus,
who chronicled Herod's deeds in great detail, did not mention this
heinous act; thus they have argued that this story was propaganda
to smear Herod. Nevertheless, there is no guarantee that Josephus
covered every ruthless deed of Herod. Merciless acts were not
unusual for Herod—a fact to which his sons can attest.

Herod killed two of his sons, Alexander and Aristobulus, in
7 B.C., largely on the basis of trumped up charges that they had
plotted their father's assassination. Another son, Antipater, was
caught planning to poison his father; he was executed five days
before Herod's own death. The annihilation of threats, real or
imagined, does not seem to have been beyond Herod, even when
his own sons were involved. Nevertheless, there is no corroborating evidence that Herod killed any children who were born under
regal stars. Thus, a neutral stance will be assumed, and the story
about the Slaughter of the Innocents will be used as an indication
that Herod was still alive two years after the birth of Jesus. That is,
the evangelist believed that Herod reacted to the threat two years
after the celestial portent. This account points to a birth for Jesus
in 6 B.C. or earlier. Figure 10 summarizes the time frame estimates
of the birth of Jesus.

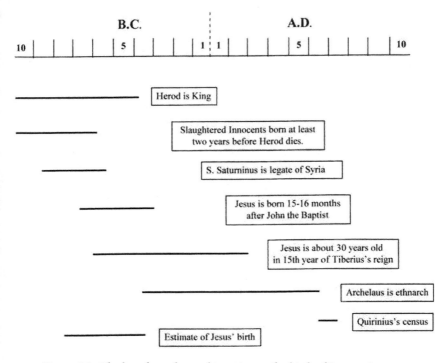

Figure 10. *The benchmarks used to estimate the birth of Jesus point to 8÷4 B.C. as the most likely period.*

From the above arguments it is reasonable to say that Jesus was born between 8 B.C. and 4 B.C., with the highest probability being that the birth was around 6 B.C.[64] When, in the next chapter, we begin our search for the Magi's star, an even larger window in time will be used to satisfy proponents of other dates and to allow some latitude for possible error.

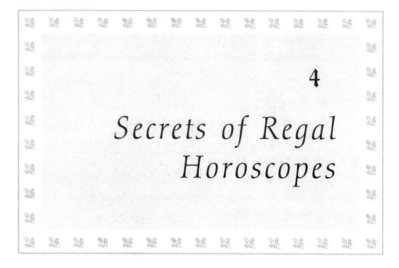

4

Secrets of Regal Horoscopes

The principles of Greek astrology concerning regal births are not intuitively obvious, and even a quick perusal of the subject reveals that Greek astrology is complex and arcane. Nevertheless, we can investigate the Magi's star by limiting the discussion to a few principles that were known as important in producing a royal horoscope.[1]

Astrologers of Roman times scrutinized natal celestial charts, which depicted the important celestial configurations when the infant was born. These charts, which we call horoscopes, consisted of two important components.[2] The first centered on the sky, specifically the signs of the zodiac and the positions of the Sun, the Moon, and the five known planets. (Astrologers used the term "planet" for the Sun, Moon, and actual planets; so shall we.) The astrologer placed the planets in the zodiac according to mathematical calculations and studied the geometric patterns formed by their positions. Triangles, squares, and hexagons, called *aspects,* were believed to produce certain conditions. Moreover, each planet had certain properties that were believed to influence the destiny of the newborn child. And other influences were thought to arise from the characteristics and properties of the zodiacal signs. Furthermore, specific sectors within the zodiacal signs, and even individual degrees, held special powers. (See Appendix B for technical details.)

The second component of the chart focused on the positions of the planets with respect to the horizon, which determined other powers and influences. (The best known of these positions is the *Ascendant,* which is essentially that part of the zodiac rising on the eastern horizon.) The entire sky above, as well as below, the horizon entered into the astrological forecast, which means that a celestial body or zodiacal sign need not be visible to play an important role in the prognosis—a fact that nearly all considerations of the Magi's star fail to take into account.

The positions of the signs of the zodiac, the planets, and the horizon were laid out in the horoscope, which more or less depicted the sky at the moment and place of birth. However, a horoscope is an abstract representation of the sky; it is not a perfect star map but more like a schematic diagram that symbolized beliefs about how the heavenly denizens determined a person's destiny. The charts from the time of the Magi consisted of a box of triangles or a tilted checkerboard diagram. Figure 11 shows part of a Greek manuscript discussing the horoscope of the emperor Hadrian. Such discussions were often forbidden because horoscopes were believed to divulge weaknesses in character, dangerous events during the lifetime, and even the life expectancy. In the hands of enemies a horoscope could become a self-fulfilling prophecy; that is, enemies could ensure the outcome of bad portents. Thus, royal horoscopes were rarely made public.

Redrawing the horoscope in figure 11 and using modern conventions and notations, produces figure 12. The view portrayed in the horoscope is facing south, and this snapshot in time shows

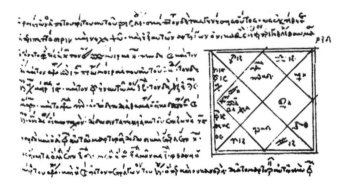

*Figure 11. Emperor Hadrian's horoscope shows how natal as-
trological charts were drawn in ancient times (2d century
A.D.). Frederick H Cramer,* Astrology in Roman Law and Politics *(Philadelphia: American Philosophical Society, 1954; reprint, Chicago: Ares Publishers, 1996), 165—With permission
of the American Philosophical Society and Ares Publishers.*

the surrounding sky: the rising eastern sky is on the left and the
setting western sky is on the right. Thus, the sky is pictured as a
giant clock slowly rotating clockwise, rising in the east and setting
in the west. The horizontal line slicing through the zodiac is the
earth's horizon, and the newborn child, Hadrian in this case, is at
the center. The dashed Midheaven line runs from the birthplace
up to the highest part of the zodiac, and extends down to the
lowest point. The tilting of the Midheaven line is explained in
Appendix C.

The zodiac and planets are arranged according to the time and
place of birth, and the positions of the planets are determined
from calculations, not observations. According to this horoscope,
Hadrian was born just before dawn: the Sun ⊙ is just about to rise
in the east. We will use Hadrian's horoscope and its astrological
analysis later to learn some of the conditions that were believed
to produce kings and emperors.

We can begin our discussion of regal horoscopes by focusing
on four conditions that, by themselves, were not always statisti-
cally rare but when considered together produced fairly unique

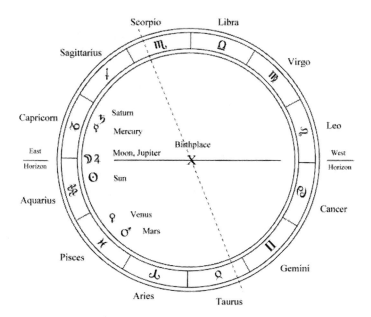

Figure 12. In this modern representation of Antigonus's horoscope for Emperor Hadrian the zodiac circles the place where Hadrian was born.

horoscopes. Moreover, they form a core group of conditions that astrologers considered to be the most important for royal births. The sources for these astrological effects are the works of Ptolemy, Dorotheus, Valens, Firmicus, and other astrologers of Roman times, who often referred specifically to several of these conditions in the horoscope of an emperor or a king. Thus, these conditions will be called the regal principles. These conditions were necessary to one degree or another to build the case for a regal horoscope. But individually they were insufficient to produce a horoscope that held royal potential, and sometimes they were augmented by other conditions. In any case, the principles will be explained as they apply to a newborn king of Judea, which will give a simplified discussion of a rather complex subject. In particular, how these principles involve Aries, the sign of the Jews, is of primary importance.

Exaltations

Many people familiar with modern astrology might expect that the domiciles (houses), zodiacal signs in which the planets hold favorable influences, would enter in regal horoscopes (see Appendix B). However, although domiciles are sometimes mentioned as additional supportive effects, they are rarely considered by themselves as indications of a regal portent. But there was another kind of house that did produce conditions conducive for a royal birth: the *exaltations,* which Greek astrologers adopted from the Babylonians. While an exaltation was a specific location (degree) along the zodiac where a planet assumed omnipotent powers, astrologers claimed that a planet was exalted simply by being in the sign of its exaltation.[3] The planet's influences were more powerful there than at any other location. Figure 13 shows the planets in their respective exaltations.[4]

Figure 13. The planets are shown in their exaltations, but the appearance of a planet anywhere in the sign was believed to bestow power to the planet.

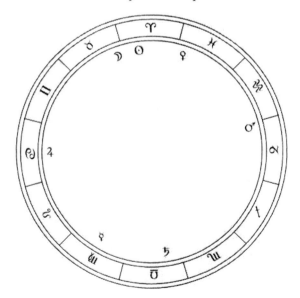

Many horoscopes have a planet in the sign of its exaltation. Nevertheless, astrologers believed that natal horoscopes involving exalted planets yielded prosperity and even had regal potential if other conditions were satisfied. However, having many exalted planets, say five, in a horoscope was considered too portentous and signified an eventual downfall—perhaps due to a terminal case of hubris.

Because Aries, the sign of Judea, is the place in which the Sun is exalted, we might very well expect that a regal horoscope for a king of Judea would have the Sun in Aries. However, because the Sun is in Aries for a month every year, in the spring, we cannot use this configuration to pinpoint a specific day for a Jewish king's birth. Moreover, an exalted Sun does not by itself point unambiguously to a regal birth in Judea. We must use the other regal principles to determine whether a specific day is truly auspicious for Judea.

Rulers of the Trines

Astrologers of Roman times paid very close attention to the power of the trines and their rulers for creating regal horoscopes. Trines are produced by any three zodiacal signs that are separated from each other by 120°. Each of the four trines was given three controlling planets called rulers (or sometimes lords; see figure 14). The presence of rulers in their respective trines was a powerful condition.

We can best understand the principle of trines by considering the trine that was of particular importance to Judea, Trine I, which has at its corners Aries ♈, Leo ♌, and Sagittarius ♐. The three planets that astrologers declared to be the rulers of this trine are the Sun ☉, Jupiter ♃, and Saturn ♄. The Sun ruled the trine during the day, and Jupiter ruled during the night.[5] Because the Magi were looking for the birth of a king of Judea they would have studied the planets present in the three signs of this trine. There are many possible configurations that place one or more of the rulers within the trine, but figure 15 shows the optimum one, in which the Sun, Jupiter, and Saturn are all in Aries. This

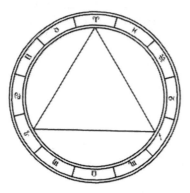

Trine I
Aries, Leo, Sagittarius
Day Rulers: Sun, Jupiter, Saturn
Night Rulers: Jupiter, Sun, Saturn

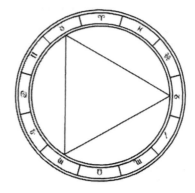

Trine II
Taurus, Virgo, Capricorn
Day Rulers: Venus, Moon, Mars
Night Rulers: Moon, Venus, Mars

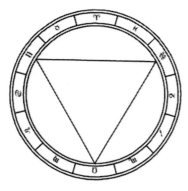

Trine III
Gemini, Libra, Aquarius
Day Rulers: Saturn, Mercury, Jupiter
Night Rulers: Mercury, Saturn, Jupiter

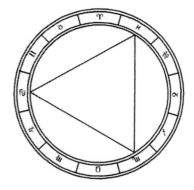

Trine IV
Cancer, Scorpius, Pisces
Day Rulers: Venus, Mars, Moon
Night Rulers: Mars, Venus, Moon

Figure 14. The four trines were assigned ruling planets, and the presence of a planet in its trine gave powerful regal significance to a horoscope.

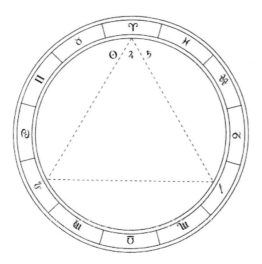

Figure 15. When all of the rulers of Trine I (Aries, Leo, Sagittarius) are in Aries ♈, attention focuses on the country governed by Aries, namely, Judea.

configuration represents an incredibly powerful portent that unambiguously points to regal blessings for Judea. A less impressive but still quite portentous arrangement would be, say, the Sun in Aries, Jupiter in Leo, and Saturn in Sagittarius.

The optimum configuration depicted in figure 15 happens approximately every sixty years. Although something that occurs every sixty years may not seem rare from a modern perspective, it would have been rare from an ancient perspective: the people of ancient times rarely lived that long, and, thus, this gathering in Aries would truly have been a once-in-a-lifetime event. Although this planetary gathering is a conjunction, it is important to realize that astrologers would have marveled at this event not because it was a conjunction but because the condition for the rulers of the trine had been superbly met.[6] Of course, the condition for the trine is also fulfilled in the more common cases when the rulers fall, together or separately, in the other two signs that make up the trine. Those instances can happen months apart for certain years, but they were less significant than events centering on Aries.

Cardinal Points

The third regal principle of ancient astrology centered on the *cardinal points* of the sky, which are four important "angles," or zones, in the sky. The astrologers often refer to these as *cardines,* and they were ranked in the following order:[7]

- Ascendant (Horoscopus)—the rising sector
- Midheaven (Medium Caelum)—the highest sector
- Descendant (Dusis)—the setting sector
- anti-Midheaven (Imum Caelum)—the lowest sector

A planet in any of these locations was believed to have extra influence on the future of the newborn infant. And the Ascendant and the Midheaven were by far the most important in producing extraordinary conditions. Thus, these two cardinal points, called "primary angles," figured prominently in regal horoscopes.

Figure 16 shows the same horoscope as figure 12, but the cardinal points have now been superposed. (As we have seen, the line connecting the Midheaven with the anti-Midheaven can be tilted. For technical details, see Appendix C). This horoscope was held to be truly remarkable because Jupiter ♃, the Moon ☽, and the Sun ☉ are in the Ascendant, which fulfilled one important regal principle. In the records of regal horoscopes, the Sun, the Moon, and Jupiter are usually located in cardinal points, especially the Ascendant and Midheaven. For example, here is an intriguing excerpt from Firmicus about the Sun in the Midheaven:

> The Sun in the Midheaven . . . *in his exaltation,* makes kings, generals, governors, consuls or proconsuls. (*Mathesis* 3.5.34)

As noted previously, when the Sun is exalted, it is in Aries, the sign of Judea. Thus, Firmicus reports that the Sun in Aries in the Midheaven is a regal portent. Of course, there were other condi-

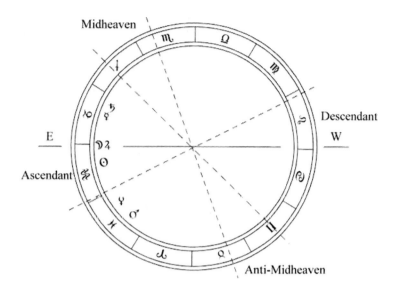

Figure 16. Superposition of the four cardinal points on Hadrian's horoscope shows that the Sun, the Moon, and Jupiter were in the Ascendant at the time of his birth.

tions that determined whether a person was destined to be a king or just a governor.

A problem with using the cardinal points is that the time of birth must be known. Therefore, this regal principle cannot be used as an explicit test for the Star of Bethlehem because the hour of Jesus' birth is unknown.

Attendant Planets

The fourth principle, which is perhaps the most subtle, describes how the five planets served as *attendants* for the Sun and Moon. Astrologers referred to attendant planets as "spearbearers" because they played a protective role, guarding the king and queen of the sky, namely, the Sun and Moon. It was believed that the good influences of the Sun and Moon could be diminished by "attacks" from badly positioned planets. Hephaestion of Thebes recorded that the rising Sun receives this protection when

it is preceded by Jupiter ♃ or Saturn ♄, that is, when the rising of
the Sun is preceded by the rising of Jupiter or Saturn. And the Moon
receives this protection when it is followed by Venus ♀ or Mars
♂. Mercury ☿ can be an attendant for either the Sun or Moon
because of the "mercurial" nature that astrologers assigned to it.[8]

Many people mistakenly believe that discussions about regal
horoscopes were strictly forbidden. True, the death penalty was
delivered for discussing a reigning emperor's horoscope, but restrictions waxed and waned over time.[9] Astrologers openly discussed theoretical regal charts or used the charts of dead people
as examples. Dorotheus of Sidon and Firmicus wrote openly about
conditions that produced regal births, and Claudius Ptolemy in
book four of his *Tetrabiblos* discussed some of the conditions for
imperial birth charts. Ptolemy's cryptic passage about attendant
planets that produce regal births follows:

> For if both luminaries are in masculine signs and either both of
> them, or even one of the two, angular, and particularly if the lumi-
> nary of the sect is also attended by the five planets, matutine to the
> sun and vespertine to the moon, the children will be kings. And if
> the attendant planets are either themselves angular or bear an as-
> pect to the superior angle, the children born will continue to be
> great, powerful, and world rulers, and they will be even more for-
> tunate if the attendant planets are in dexter aspect to the superior
> angles. (*Tetrabiblos* 4.3)

Such opaque and arcane writing could very well be a reason that
people have advocated much simpler explanations for the Star of
Bethlehem. Anyone can understand a comet, but attention spans
are challenged by ancient astrological terminology. Here is a simplified explanation of this terribly complex but very important
passage that describes the optimum conditions for a regal birth:

- The Sun or the Moon is in *Aries,* Gemini, Leo, Libra, Sagit-
  tarius, or Aquarius (the so-called masculine signs).
- The Sun or the Moon is in the Ascendant, the Midheaven, the
  Descendant, or the anti-Midheaven.

- In a day birth, Saturn, Jupiter, or Mercury rises before the Sun as an attendant planet.
- In a night birth, Mars, Venus, or Mercury sets after the Moon as an attendant planet.
- The attendant planets should be in the Ascendant or the Midheaven, or in a trine that includes the Ascendant or Midheaven.[10]

At first Ptolemy's conditions for attendance appear easy to fulfill; however, most if not all of these conditions must be met for an auspicious chart. Some horoscopes have only one attendant planet, whereas others have several. Furthermore, one horoscope may have the Sun in the anti-Midheaven, but another may have the Sun in the more important Ascendant. An astrologer could rule out the less impressive first case and declare that the weaker horoscope was fitting for a notable governor perhaps, but not for a king. Thus, for a horoscope to be undeniably suited for a royal birth it must have a strong set of conditions for attendance.

What would be a strong case of attendance that points unambiguously to a regal birth in Judea? According to the first condition either the Sun or Moon should be in Aries, and it should be either preceded or followed by its respective attendant planets. To keep the explanation simple, the case for the Sun alone is shown in figure 17. In this example of regal attendance the time is just before local noon, when the Sun ☉ is approaching the Midheaven. Jupiter ♃ and Saturn ♄ precede the Sun. Note how this alignment simultaneously fulfills all the conditions for the rulers of Aries' trine (see figures 14 and 15); that is, the Sun, Jupiter, and Saturn rule the trine of Aries. Furthermore, the Sun is exalted in Aries and in a primary cardine, the Midheaven. This portentous configuration by itself serves as an unambiguous indication of a regal birth in Judea.

Firmicus also discussed regal horoscopes in his *Mathesis,* in a chapter called "Royal Charts." He repeated essentially the same arguments that Ptolemy gave regarding the importance of the Sun and Moon, the attendant planets, the Ascendant and Midheaven, and so forth. More important, he added that there was one planet

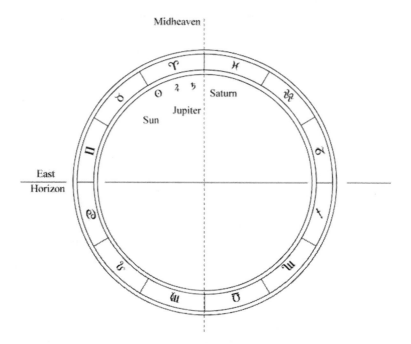

Figure 17. A regal portent for Judea is formed when two of the rulers of Trine I, Jupiter and Saturn, also serve as attendant planets for the third ruler, the exalted Sun in Aries ♈.

that could offset any negative influences that a potentially hostile planet, Mars or Saturn, might introduce if it was not properly aligned in the procession of planets:

> If the influence of Jupiter is added, imperial command is entrusted to them [people born with this horoscope]. (*Mathesis* 7.22.1)

This remark about Jupiter's power to bestow the emperorship has important implications concerning the Magi's star.

Regal Stars and Planets

Some people may see the frequency with which Jupiter and Saturn are mentioned in the regal principles as confirmation of

Isaac Abarbanel's ideas about portents for the advent of the Messiah: Abarbanel felt that conjunctions between Jupiter and Saturn were related to the rising of great leaders among the Jews. However, when his theory about Pisces as the sign of the Jews is used, the importance of Jupiter and Saturn vanishes because Mars and Venus are the rulers of the trine of Pisces (see figure 14). Thus, Abarbanel's ideas do not agree with ancient astrological beliefs, but it is possible that he heard about these conjunctions from astrological sources and applied them to the wrong sign—an intriguing thought.[11]

All in all, the four regal principles illustrate how intricate, arcane rules were formed from simple, visually unimpressive celestial alignments. Nevertheless, these conditions worked synergistically to form unusual horoscopes. Although the tenets of Greek astrology dictated that several conditions were needed to produce a regal portent, the biblical account in Matthew speaks of "his star," not "his stars." Thus, the next task is to find a situation when the regal principles came together to produce a magnificent regal portent in which a star played a central, dominant role.

Could the Magi's star have been an ordinary star with a special astrological meaning? There were special "bright stars" that figured in astrological prognostications. And an anonymous astrologer of A.D. 379 gave an extensive treatise on the subject of "bright stars," and other astrologers such as Firmicus discussed their importance as well.[12] About thirty stars in the constellations were assigned astrological powers. And a birth that occurred when the Moon was in conjunction with a bright star was believed to be important. But, if the bright star were a special one with regal powers, the lunar conjunction could reveal a royal birth. The most famous of these regal bright stars is Regulus, the brightest star in Leo, the Lion, a star that is often referred to as the "heart of the lion." Ptolemy called this star Βασιλίσκος (*Basiliskos* = the king).

The role that bright stars performed in an astrological prognostication was to increase the influence of the other positive configurations. A fine example of this concept can be found in the famous royal horoscope of Antiochus I of Commagene. On his tomb at Nemrud Daği in modern Turkey, there is an astrological diagram, which has been interpreted to signify the royal corona- tion of Antiochus or, as archeologist Donald Sanders believes, the regal foundation of Nemrud Daği. On the stone relief, the Moon, Mercury, Mars, and Jupiter are in the zodiacal sign Leo, the Lion. Over the heart of the lion is the bright star Regulus.[13] This regal star is shown prominently in the arms of a crescent moon, a con- figuration that symbolizes majesty or sovereignty (see figure 18).

Nowadays, the star and crescent symbol is associated with the Muslim nation; however, the symbol was used on ancient coinage almost a thousand years before the time of Muhammad. Many scholars maintain that the Western interpretation of the symbol arose from Babylonian mythology in which the juxtaposition of Sin (moon god, father of time) and Shamash (supreme ruling sun god, judge of heaven and earth) was a metaphor for the cosmic powers given to the Babylonian king to rule.[14] However, if we consult Greek astrology, we find that an important astrological

Figure 18. The Regal Lion Horoscope of Antiochus I at Nemrud Daği shows the regal bright star Regulus in the arms of the crescent moon— a portent signifying majesty. Photo courtesy of Donald H. Sanders.

portent was symbolized by the star and crescent: an occultation (the obscuration of one celestial body by another) of a "bright star" by the Moon.

> If, then, you should find at a birth that the Moon is coming along-side one of the bright and notable stars, meaning that it has the same number of degrees as they do [same longitude: position along the zodiacal sign], and especially if the Moon should run according to the same latitude [same distance from the ecliptic] as does also the bright star which is near it in degrees, they make the births great, bright, most notable, and well off." (Anonymous of 379 *Treatise on the Bright Fixed Stars* 1.1)

Astrologers monitored four regal bright stars in particular for conjunctions with the Moon. These bright stars, Aldebaran in Taurus, Regulus in Leo, Antares in Scorpius, and Fomalhaut in Piscis Austrinus, were also important when they were rising, that is, in the Ascendant.

That none of these regal bright stars is in Aries means that it is not obvious that they played a role in forming the Star of Bethlehem.[15] However, Firmicus tells us about a less important bright star in Aries, Hamal, which is the brightest star in the constellation.

> A bright star [is] in the eleventh degree of Aries. Whoever has the Ascendant in this star, if Jupiter is with it or in trine aspect to it, will be a great, powerful leader, a friend of kings, holding important land. He will die at the appointed time. (*Mathesis* 8.31.2)

This bright star in Aries did not confer kingship, but it did have the potential to make the infant a leader. However, the Magi announced to Herod that a king had been born, not a "powerful leader, a friend of kings." Thus, there is no strong case that this bright star in Aries was the Star of Bethlehem, although its presence may have helped to enhance regal conditions involving Aries. By coincidence, Firmicus mentioned in this quote the influence of Jupiter in regal charts, which leads to the next logical candidate for the Magi's star, namely, a single planet, specifically Jupiter.

In Greek astrology the planets, Sun, and Moon were called "stars," and they were assigned the powers that Greek mythology bestowed upon them. Here Vettius Valens tells us about those properties.

> Each star is the ruler of its own essence. . . . The Moon is set down as ruler of foresight, the Sun of light, Kronos [Saturn] of ignorance and necessity, Zeus [Jupiter] of opinion and crowns of office and will, the star of Ares [Mars] of action and troubles, the star of Aphrodite [Venus] of love and desire and beauty, the star of Hermes [Mercury] of law and custom and fidelity. (*Anthology* 1.1)

In Babylonian and Assyrian mythology, the god Marduk was the savior of the world; Babylonian astrologers connected his star to the destiny of the king.[16] Building upon this myth, the Greeks related Marduk to Zeus, who saved the world from the ravages of evil Typhon. Zeus, known to the Romans as Jupiter, was the king of the pantheon. In Greek astrology the star of Zeus was believed to manifest his divine powers in matters involving fame, prosperity, happiness, politics, greatness, and kingship; thus, the planet Jupiter figured prominently in offering the royal purple.

In recent years, there have been more theories about the Star of Bethlehem focusing on Jupiter.[17] Although its appearance in the sky is not always dazzling, the astrologers of antiquity gave ample evidence for its regal importance. For instance, Dorotheus of Sidon says that a great person is born under Jupiter's influence:

> If Jupiter is in the Midheaven, then the infant will be far out in his fame, noble, having good fortune and children, especially in a day birth. (*Pentateuch* 2.24.6)

Firmicus also says that if Jupiter rises before the Sun that is *in Aries,* we have imperial conditions:

If the Sun is on the Ascendant . . . in exaltation [in Aries], this will make governors of the most important provinces or states. . . . If Jupiter in this combination protects [as an attendant] either the Sun or the Moon or both, this makes most powerful emperors, just and fortunate, universally regarded with terror. (*Mathesis* 3.13.10)

The importance of Jupiter in regal horoscopes was believed to be magnified when it was in conjunction with the Moon, which is a key to finding a special horoscope.

A   R E G A L   L U N A R   C O N J U N C T I O N

The Moon and Jupiter played important roles in imperial horoscopes. Antigonus of Nicaea, a physician-astrologer of the second century A.D, made an analysis of the important interaction between these two celestial bodies in Emperor Hadrian's horoscope (see figure 12).

That he was honored and received the *proskynesis* [worship] from all men was due to the fact that Jupiter was in epicentric attendance on the Sun [rose immediately before the Sun]. For a planet that attends in this manner the Sun and the Moon has the effect that a man is highly esteemed by his equals or superiors, and has attendants and receives the *proskynesis*. Also, he received the beneficial qualities from the aforesaid position of Jupiter.[18]

In any case, the significant information from Antigonus is that Jupiter was the central reason for Hadrian's imperial destiny. And, in particular, the powers of Jupiter were amplified by the proximity of the Moon and the fact that both were in the Ascendant. Antigonus's conclusion about Jupiter and the Moon is in full agreement with that of other astrologers:

If she [the Moon] is found on the first angles [the most important cardinal points], this is in the Midheaven, or the Ascendant, and is in aspect to Jupiter, she will make powerful emperors or kings or governors with world-wide power. (Firmicus *Mathesis* 5.6.6)

Firmicus confirms the significance of a conjunction between the Moon and Jupiter. Furthermore, he predicts infinite rewards if the Moon and Jupiter are very close:

Jupiter and the Moon in the same sign indicate the greatest good fortune, especially if they are in the same degree. For then they bestow infinite riches and marks of prosperity that the natives are always superior to their parents. (*Mathesis* 6.23.7)

This prediction is reinforced by Vettius Valens, who recorded that the Moon in good aspect (in conjunction or trine) with Jupiter produces "notable leaders."

Zeus [Jupiter] and the Moon, then are good, acquisitive, producing masters of ornament and of bodies, and furnishing notable leaders . . . , and those who are deemed worthy of gifts or honors. (*Anthology* 1.19)

This passage rings with special meaning because the Magi brought gifts to honor the newborn child.

Some might imagine a conjunction of the Moon and Venus had regal implications.[19] Indeed, such conjunctions in the evening or morning skies are striking and beautiful. Furthermore, the well-known Babylonian interest in Venus (Ishtar, the queen of the sky) strengthens the case for a close lunar conjunction with Venus. Nevertheless, the record of Greek astrology does not support those notions. Lunar conjunctions with Venus were quite favorable, but they did not have the regal significance that Jupiter invariably commanded. Here is what Firmicus has to say both about the waxing Moon visible in the evening sky after sunset as it comes into conjunction with Jupiter and about a similar conjunction with Venus:

If the waxing Moon is in aspect to Jupiter or is moving toward him, the natives will be fortunate, famous, and rich; masters of many great estates and wide possessions. (*Mathesis* 4.3.1)

If the waxing Moon is moving toward Venus or coming into aspect with her, the *parents* will be noble, raised up to highest power. But the natives will be separated from the parents and suffer as wretched orphans. They will, however, attain high position, be well-known, successful in all their activities, and easily attain their goals. (*Mathesis* 4.6.1)

So, a lunar conjunction with Venus was not seen as significant as that with Jupiter.

Which conjunctions or aspects were believed to produce imperial conditions? According to Vettius Valens's *Anthology,* regal portents were produced most often by the following pairs of celestial bodies: Sun-Jupiter, Moon-Jupiter, and Sun-Moon.[20] Thus, modern impressions about the striking spectacle that Venus makes in the sky do not agree with ancient perceptions. This reinforces the importance of thinking like an astrologer during King Herod's reign, not like an archaic Babylonian astrologer or a modern person.

The astrological record shows how the Moon's interactions, particularly with Jupiter, were scrutinized for indications of a royal birth. In Emperor Hadrian's horoscope the Moon was said to be in the "same degree" as Jupiter, a condition that played a commanding role according to the astrologer Antigonus. Although the Moon passes Jupiter each month to produce a conjunction, Antigonus pointed to other astrological conditions that augmented the conjunction and gave Hadrian the *proskynesis,* the worship of the people.[21]

Although lunar conjunctions were commonplace, there was a unique one that had the attention of astrologers. Historian Otto Neugebauer found that the records of Greek astrology indicate that many astrologers were especially interested in close conjunctions involving the Moon and in occultations in particular.[22] An occultation, the closest possible conjunction, occurs when the Moon completely obscures a planet. Technically, this could be thought of as an eclipse, but astronomers and astrologers reserved the word "eclipse" for a situation in which the Moon obscures the Sun or the Moon passes into the earth's shadow. According to astrologers, occultation maximizes the interaction between the two bodies and intensifies their astrological effects.[23]

Not only does a lunar occultation of Jupiter have regal implications in a horoscope, but it also provides a simple celestial event that points to a specific day when regal conditions were exceptional. If a lunar occultation of Jupiter happened in Aries, it could have signaled to the Magi that a king had been born in Judea. Of course, such an event cannot by itself produce a regal horoscope. The significance of the lunar occultation must be sustained by other astrological conditions, namely, the regal principles. Furthermore, the account in Matthew must be consulted to determine whether there is supporting evidence that such a finding fits the description of the Magi's star. In any case, our search for the Star of Bethlehem will now focus on a lunar occultation of Jupiter, the regal star of Zeus.

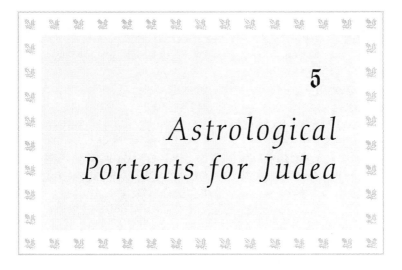

5

Astrological Portents for Judea

Raised in a modern, technological society, people today are prone to equating important astrological portents with visually spectacular events. Thus, the most popular theories concerning the Star of Bethlehem are those that offer the grandest celestial spectacle, such as a comet or a supernova. However, the biblical account of the star contains a clue that points to the flaw in this modern perception: the people of Judea did not know about the Magi's star.

Rather convoluted scenarios have been contrived to explain why only the Magi saw the star. Some scenarios have the star playing a magical game of hide-and-seek. After the Magi saw the star appear in the east, they departed on their journey to Jerusalem. When they arrived, the star disappeared and could not be seen by Herod and the people of Jerusalem. After their audience with King Herod, the Magi saw the grand star again and praised it. However, an astrological interpretation of the Star of Bethlehem gives a simpler and more believable explanation of the biblical account: the people of Jerusalem simply did not recognize the significance of the star.

When my research on the coins of Antioch found that Aries, the Ram, was the sign governing Herod's kingdom, I formulated a theory around a specific regal portent involving the planet Jupiter, with a lunar occultation of the planet in Aries as the central element forming regal conditions. Such a portent in Aries would have pointed like an astrological road sign to Jerusalem, the governing capital of the lands under Aries' control.

In chapter 3, the time of Jesus' birth was estimated to lie most likely in the period from 8 to 4 B.C. If this interval is enlarged to 10 B.C. to A.D. 5, it satisfies almost all estimates of Jesus' birth.[1] Now these years must be examined for any lunar occultations of Jupiter in Aries. Finding such an occultation might well point to the birth of a king of the Jews.

Amazingly, not one but two lunar occultations occurred in Aries during this period, and both involved Jupiter. Most significant, they happened in 6 B.C., during the reign of King Herod, in the middle of the estimated time frame for Jesus' birth. On March 20, 6 B.C., just before sunset in Judea, the Moon occulted Jupiter while in Aries. The occultation ended half an hour later almost on the western horizon. Jupiter was too close to the Sun to be visible. Although astrologers did not have the accuracy of modern methods, they did have the mathematical means to know that the paths of the Moon and Jupiter had crossed in Aries to produce a close conjunction, perhaps even an occultation; and modern calculations verify that this occultation did indeed occur.[2]

The spring of that year was conducive for close lunar conjunctions with Jupiter, and a second occultation of Jupiter happened the following month. On April 17, 6 B.C., the Moon, having circled the sky, returned to Aries and again occulted Jupiter, a little after local noon, when Jupiter was still too close to the Sun to be seen. Although these occultations are intriguing, further evidence is necessary if we are to convincingly relate one or the other to the Magi's star.

Now we must consult the biblical account for any evidence that confirms either of these celestial events as the Magi's star. Here is the passage describing the star:

> When they had heard the king, they departed; and, lo, the star, which they saw in the east, went before them till it came and stood over where the young child was. When they saw the star, they rejoiced with exceeding great joy. (Matt. 2:9–10)

Biblical scholars have noted the mysteriousness of the star described in Matthew. The star ostensibly appears in the east and moves before the Magi as they travel from the east, a seeming physical contradiction. Then, the star amazingly stands above the child, an apparently unusual celestial event. There is no reason to doubt that the Magi came from east of Judea, where astrology was practiced. But the puzzle is about a star "in the east," a phrase that in a literal translation means to some people that the star was behind the Magi as they traveled westward to Judea. Other people believe that the Magi literally followed the star that "went before" them. All of this confusion is the result of thinking like a modern person and not like an astrologer of antiquity.

The phrase "in the east" is a literal translation of the Greek phrase ἐν τῇ ἀνατολῇ (*en te anatole*), which actually means "at the rising." But more important, in Greek astrology it means specifically that a planet rises before the Sun as a morning star; that is, it undergoes a heliacal rising. Thus, although *en te anatole* translates literally as "in the east," in the parlance of astrologers it really means "at the *heliacal* rising," or "at the morning appearance." The person recounting the story of the Magi's star to the evangelist must have realized the significance of its being in the east because Matthew refers to this condition twice. From an astrological perspective, it was extremely significant that the Magi's star was a morning star.

Heliacal risings have figured heavily in the history of astronomy and astrology. To astrologers the emergence of a planet from the burning heat of the Sun meant a new birth of cosmic power—an idea stemming from Stoic beliefs about the importance of cyclical deaths and rebirths: *ekpyrosis* and *palingenesis,* a fiery death followed by rebirth. And as a planet moved away from the Sun with each day, the planet's influences were believed to be manifested powerfully. According to astrologers a planet's heliacal rising— along with the subsequent phase, in which it appeared as a morning star—was the most significant period in its procession around the sky, the period in which its astrological influences were greatest.

Calculating when a planet will have a heliacal rising, is difficult, and the calculation was thoroughly analyzed by Claudius Ptolemy in the *Almagest* (The Greatest). He demonstrated that the time of a heliacal rising depended on the planet's brightness, the position of the rising Sun below the eastern horizon, and the latitude of the observer. His calculations showed that for the springtime, when the Sun is in Aries, Jupiter must be at least twenty degrees from the Sun to be visible from Alexandria, Egypt. When the Sun is in Libra, in the fall, the angle decreases to about ten degrees. (For reference recall that the Moon and Sun are one-half degree in diameter, and a zodiacal sign is thirty degrees wide.) But this mathematical determination of first visibility in the dawn sky was not used in astrology.[3]

The astrologers envisioned that surrounding the Sun was an "arc of combustion" within which a planet's power was destroyed by the searing rays of the Sun. And when it emerged from this region, great powers burst forth from the planet. Thus, astrologers adopted a standard, fixed distance from the Sun for each planet's heliacal rising, a distance that did not require calculation of a planet's actual visibility. Each planet had its own angular distance of safety from the Sun, and that angle applied to all risings during the year. Ptolemy and other astrologers tell us that Jupiter, for example, emerged from the Sun's fire at twelve degrees.[4]

This belief in the burning power of the Sun is emphasized by Firmicus. He discusses the power of the Sun to suppress the influences of Jupiter. Moreover, the astrological significance of Jupiter is greater if it is in the same sign as the Sun:

> If Jupiter and the Sun are in conjunction and Jupiter is hidden by
> the Sun's rays, this diminishes the whole fortune. But if Jupiter,

freed from the Sun's rays and in a morning rising, is together with the Sun in the same sign, this indicates high honor and position. (*Mathesis* 6.23.3)

When did Jupiter have a heliacal rising in Aries during the time frame of interest, 10 B.C. to A.D. 5? *Only once,* on April 17, 6 B.C., which coincides exactly with the second lunar occultation of Jupiter! The first lunar occultation with Jupiter on March 20, 6 B.C., occurred close to the Sun, or as an astrologer would have said "under the searing rays of the Sun." Thus, the first occultation can be disregarded in our search for the Star of Bethlehem because its influences would have been weakened by the Sun. But the lunar occultation on April 17, 6 B.C., happened as Jupiter emerged in the east.

In his analysis of Emperor Hadrian's horoscope, Antigonus of Nicaea recorded how portentous it was for the emperor that Jupiter was going to rise in the east seven days after its close conjunction with the Moon (see Appendix D). However, on April 17, 6 B.C., the conjunction and the heliacal rising occurred simultaneously. From an astrologer's perspective, this perfect timing must have been truly stupendous. But most significant, this event happened in Aries, where the Sun was exalted. All of this pointed to the birth of a king of Judea.

What else can we find in the biblical account that points to the conjunction of April 17 as the Star of Bethlehem? The account in Matthew says that the Magi's star "went before" and then stopped and "stood over the place where the young child was." As we have seen, some have interpreted this to mean that the star disappeared after being in the east, remaining invisible to King Herod and the people of Jerusalem and manifesting itself again after the Magi left Herod in search of the child. Actually, Jupiter never disappeared after its heliacal rising, but its regal astrological significance did change over the months of 6 B.C. as it moved from the dawn sky into the night sky.

Table 5.1 gives the a literal translation of the verse in Matthew that describes the behavior of the star after the Magi departed from their audience with Herod. Ivor Bulmer-Thomas has clarified the strange movements of the Magi's star.[5] Processes like "in

Table 5.1. Greek Text of Matthew 2:9

καὶ	ἰδοὺ	ὁ ἀστὴρ,	ὃν	εἶδο	ἐν
and	behold	the star	which	they saw	in
τῇ	ἀνατολῇ,	προῆγεν	αὐτούς,	ἕως	ἐλθὼν
the	east,	went before	them,	until	having come
ἐστάθη	ἐπάνω	οὗ	ἦν	τὸ	παιδίον.
it stood	over (the place)	where	was	the	child.

SOURCE: *Robert K. Brown and Philip W. Comfort,* The New Greek-Enbglish Interlinear New Testament, *ed. J. D. Douglas (Wheaton, Ill.: Tyndale House, 1990), 5.*

the east," "went before" and "stood above" are translations of astrological terms. The word προῆγεν (*proegen,* "went before") is related to the astrological term προηγήσεις (*proegeseis*), which indeed means "to go before" or, more precisely, "to go in the same direction as the sky moves." A planet goes in the same direction as the sky when it reverses its eastward motion through the zodiac and proceeds in the same westward direction in which the sky rotates. The ancient Greeks perceived the "normal" direction for a planet as the direction of the sky's movement. Today, however, astronomers reverse this concept: they think of movement in the direction of the sky as backwards movement and call it retrograde motion.

Retrograde motion occurs because the earth is closer to the Sun than the outer planets and thus orbits the Sun more quickly.[6] Saturn, Jupiter, and Mars, along with the Moon and Sun, normally travel slowly along the zodiac toward the east, in the opposite direction of the rotating sky. Sometimes, however, these planets appear to stop for a couple of weeks or so, then reverse direction. Figure 19 shows how this happens. At position 1 Jupiter appears to be moving normally to the east. At position 2 the motion of Jupiter relative to the background stars appears to cease, and the planet reaches a "station," or stationary point. Then, at position 3, the planet appears to drift backwards. During this stage, the Greek astronomers and astrologers said that the planet moved "forward," or "went before," the background stars, namely, in the direction that the sky moves. At position 4, Jupiter halts a second

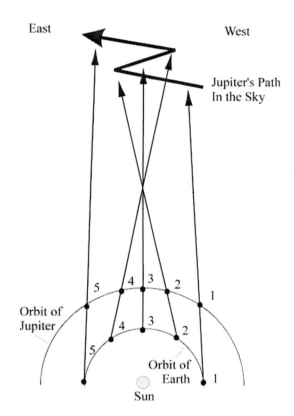

Figure 19. Retrograde motion and stationing are the apparent movements of a planet seen from the moving earth.

time at another station before resuming its usual eastward motion among the stars of the zodiac at position 5.

This phenomenon can be experienced from an automobile passing a slower vehicle against a background of distant trees. From behind, the slower vehicle is seen moving forward relative to the trees. Just as the slower vehicle is passed, it appears briefly to go backwards relative to the distant trees. Had the ancient astronomers accepted the idea that the earth traveled around the Sun, they would have understood that stationing and retrograde motion are illusions due not to some intrinsic planetary power but to the earth's motion.

Great confusion has also been created by the report that the star "stood over." The Greek word ἐπάνω (*epano*), which can mean "over" in some circumstances, is typically interpreted in Ptolemy's works as meaning "above," that is, "in the sky."[7] But biblical scholars have translated it literally, as "over." If we interpret this phrase in astrological terms, it means that the planet became stationary in the sky, which is, by definition, above the child. All in all, the passage in Matthew has strong astrological meanings, which the evangelist described in simplified, literal terms. Nevertheless, the true meaning of the message is still interpretable.

What did the ancient astrologers have to say about the significance of retrograde motion and stationing? Ptolemy says that the astrological effect of planets could be intensified when they underwent heliacal risings or became stationary.

> For planets when they are rising or stationary produce intensifications in the events. (*Tetrabiblos* 2.6)

As it turns out, the astrological importance of stations explains why they would have been mentioned in the account about the Magi. Here, Ptolemy describes the effects of the phases of planetary motion on the bodily form and temperament of a newborn child:

> Again, generally, when the planets are morning stars and make an appearance [undergo heliacal rising], they make the body large; at their first station, powerful and muscular; when they are moving forward [προηγούμενοι = *proegoumenoi*], not well-proportioned; at their second station, rather weak; and at setting, entirely without repute but able to bear hardship and oppression. (*Tetrabiblos* 3.11)

That προηγούμενοι (*proegoumenoi*), the word Ptolemy uses, is essentially the same word that Matthew uses corroborates the idea that the biblical account was applying technical astrological language that has since lost its meaning. Thus, the correct interpretation of the Magi's report is that the star was moving forward against the background stars, not in front of the Magi as they traveled from the east.

Ptolemy also repeatedly emphasizes how stationary planets affect the meaning of major astrological conditions. Here, he tells us that the planet influences the country controlled by the sign of the zodiac in which the planet stands still:

> Of the prediction itself, one portion is regional; therein we must foresee for what countries or cities there is significance in the various eclipses or in the regular stations of the planets, that is, of Saturn, Jupiter, and Mars, whenever they halt, for then they are significant. (*Tetrabiblos* 2.4)

According to Ptolemy, Jupiter's stationary point in Aries bestows special blessings on Herod's kingdom.

Now that we have reinterpreted the passage in Matthew in astrological terms, how does it fit with the actual astronomical events of April 17, 6 B.C.? "Went before" and "stood over" exactly describe what Jupiter did after being "in the east." Figure 20 illustrates several important points regarding the movement of Jupiter in the sign of Aries. First, according to ancient conventions the zodiacal sign does not necessarily conform to the constellation; the dashed outline of the sign of Aries extends into the constellation Pisces and maybe Taurus. Superposed on this star chart are the paths that Jupiter and Saturn traveled during 6 B.C. Both moved eastward (to the left on the chart) from Pisces into Aries early in the year. On April 17 Jupiter emerged in the east and was occulted by the Moon. The positions of the planets on that day are shown in the figure. On the far left, Mars ♂ is in Taurus. The Sun ☉, Jupiter ♃, the Moon ☽, and Saturn ♄ are in Aries; and Venus ♀ is in Pisces. (Mercury is out of view, in Taurus.) Of course, most of these planets were too close to the Sun to be visible on that day, but as the Sun left Aries and moved into Taurus in the following days, Jupiter became visible and slowly drifted eastward (its usual direction) through the stars of Aries. Jupiter left Aries and entered Taurus on about June 20 and then stopped at its first station on August 23. The regal planet seemed to stand motionless in Taurus for approximately a week as the earth, in its faster orbit around the Sun, started to catch up to Jupiter.

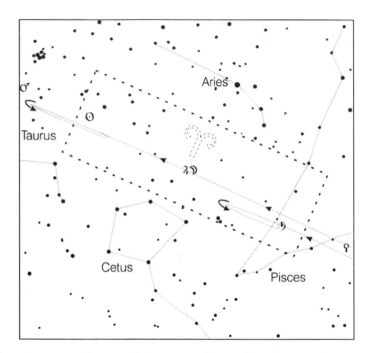

Figure 20. The retrograde motion of Jupiter ♃ and Saturn ♄ over the year 6 B.C. is superposed on the positions of the planets for April 17, 6 B.C.

Jupiter, still visible in the night sky, reversed direction and proceeded westward ("went before," in agreement with the biblical account), re-entering Aries on October 27, 6 B.C. During that time, the earth was passing Jupiter, which produced the illusion that the distant planet was going in the opposite direction against the background stars.

Jupiter then stood still a second time, for a week or so centering on December 19 in Aries—the zodiacal sign representing the place of the newborn king's birth. Jupiter finally resumed its usual eastward motion against the background stars and left Aries early in 5 B.C. Jupiter would not undergo another heliacal rising in Aries for a dozen years.

Astrologers would have noted the second stationing of Jupiter in Aries because the principle of trine rulers had been fulfilled. Figure 21 shows that during the period from November 19 to December 19, 6 B.C., the Sun passed through Sagittarius while Jupiter and Saturn were still in Aries, and thus the trine of Aries, Leo, and Sagittarius was occupied by its three rulers. However, this trine aspect was a mere echo of the far more important conditions that existed a few months earlier on April 17, 6 B.C. In November and December, Jupiter was not in morning rising and the Sun was not in its exaltation, that is, in Aries. That Jupiter did undergo retrograde motion and reach its station during this trine aspect intensified the other conditions, according to Ptolemy.[8] Thus December 19 held a secondary regal aspect confirming the birth of a Judean king.

Perhaps a simple paraphrasing of the important passage in Matthew is needed—a paraphrasing that reflects what an astrologer might have said to the evangelist.

Figure 21. A secondary regal portent formed when Jupiter underwent retrograde motion and became stationary in Aries; that is, the rulers of the trines were established again with Saturn still in Aries and the Sun in Sagittarius.

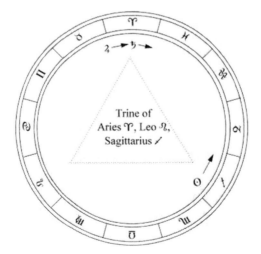

> And behold the planet which they had seen at its heliacal rising
> went retrograde and became stationary above in the sky (which
> showed) where the child was.

Matthew's reports of these planetary movements—heliacal rising, retrograde motion, and stationing—indicate that the Magi
continued to monitor the motion of Jupiter after the heliacal rising,
perhaps in search of other regal conditions as Jupiter traversed
Aries. Nevertheless, the most important astrological condition in
the biblical account occurred not when the star "went before" or
"stood over" but when it was "in the east." The biblical account
mentions "in the east" twice and for good reason: a heliacal rising
was the most portentous time for a planet such as Jupiter.

We must focus then on April 17, 6 B.C., when Jupiter, the regal
planet primarily responsible for bestowing kingship, emerged
from the debilitating rays of the Sun and manifested its astrological powers. This important transition for Jupiter was believed to
be amplified even more by the Moon's close proximity to the
planet. Surely this combination of conditions conveyed the message that a king had been born in the lands ruled by Aries. But
these conditions alone were not the reason that attention was
drawn to the sky of that day. Vettius Valens explains that not only
must the rulers of the trine be well positioned but so must the
other beneficial planets: all aspects centering on Aries must be
examined.

> *Aries:* When the rulers are well situated and testified to by benefics,
> the natives become kingly, powerful, and have the power of life
> and death. (*Anthology* 1.2)

Thus, the question before us is whether the astrological aspects
were "well situated," revealing a royal birth.

THE GREAT MESSIANIC PORTENT

If the heliacal rising and lunar occultation of Jupiter constitute the
Star of Bethlehem, and are not just fortuitous coincidences, a

horoscope drawn for April 17, 6 B.C., must prove to be significant according to the regal principles of Greek astrology. Furthermore, according to the royal horoscopes the preeminence of Jupiter, the regal planet, must be unmistakable. And most important, the horoscope must be so incredibly portentous that it points unquestionably to a regal birth in Judea.

As it turns out, selecting the hour of the birth, which is needed to analyze the influences of the cardinal points (Ascendant, Midheaven, and so on), is not important for that day. That is, any horoscope drawn for that day depicts incredible royal portents, although some hours are, of course, much better than others. According to the astrological commentaries on imperial horoscopes, two important times are sunrise and the time when Jupiter is in the Midheaven, and horoscopes for these times are shown in figures 22 and 23. Both horoscopes are striking, even from an astronomical perspective: all of the planets cluster in and around Aries.

If we apply the regal principles discussed in chapter 4 to the horoscopes shown in figures 22 and 23, we find that the day was

Figure 22. Sunrise for April 17, 6 B.C.

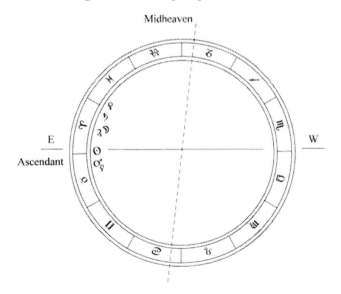

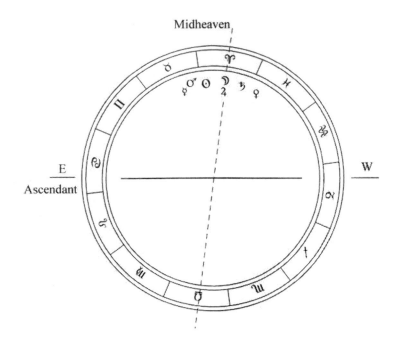

Figure 23. Jupiter at the Midheaven on April 17, 6 B.C.

truly astounding. And special note should be made of the number of important aspects and conditions that involve Aries, which would have directed an astrologer's thoughts to King Herod's realm:

- Exaltations: In both charts the Sun is in Aries. The exaltation of the all-important Sun ☉ in Aries draws attention to that sign. A beneficial planet, Venus ♀, is exalted in Pisces ♓. Two exalted planets give regal potential, but more important, the Sun's exaltation in Aries emphasizes Judea.
- Rulers of the trines: Both charts are for the day, a time that bestows amazingly strong portents. The Sun, the ruler of daytime, is in Aries and, therefore, is the primary ruler of its trine (Aries, Leo, and Sagittarius). Jupiter ♃, a co-ruler of the same trine, is also present in Aries; and, incredibly, so is the third

ruler of the trine, Saturn ♄! This truly auspicious configuration signifies the importance of Aries. If a night chart (nocturnal birth) were drawn, Jupiter, the regal planet, becomes the primary ruler, and the co-rulers are the Sun and Saturn. Both cases, day and night, produce stupendous horoscopes.

- Cardinal points: The time of Jesus' birth is unknown; hence, no conclusions can be drawn about whether the positions of the planets relative to the cardinal points made his horoscope even more important. The horoscopes in figures 22 and 23 show two important times, sunrise and the time when Jupiter (as well as the Sun) is in the Midheaven, times that enhance the other principles.[9] (For the influences of other configurations see Appendix C.)

- Attendance: Saturn and Jupiter, "spear-bearers" for the Sun, rise immediately before the Sun in truly perfect fulfillment of the requirements for attendance. In fact, their preceding the Sun is a textbook example of regal attendance. This horoscope is similar to that of Hadrian; in both horoscopes the Moon would also have been considered a protective attendant for the Sun because it rises just before the Sun. In addition, Mars and Mercury follow the Moon and serve as its attendants. Above all, the gathering of the Sun, the Moon, Jupiter, and Saturn in Aries on April 17 gives regal conditions even more portentous than those in Hadrian's or Augustus Caesar's horoscope (see Appendix D).

- Beneficent planets: In addition to being a morning star, Jupiter is also a co-ruler of Aries, an attendant for the Sun, and in close conjunction with (occulted by) the Moon. These three conditions alone set the stage for a regal birth: Jupiter's influence in the charts for April 17 is vastly superior to its influence in Hadrian's chart, which depended upon Jupiter for its regal conditions. Jupiter is, indeed, the central "star" for April 17, 6 B.C. The other beneficial planet, Venus, is in Pisces and the day ruler of its own trine (Cancer ♋, Scorpius ♏, Pisces ♓)—another very fortunate condition that reinforces the regal nature of that day.

- Maleficent planets: Mars ♂ is present in the trine in which it is a co-ruler (Taurus ♉, Virgo ♍, Capricorn ♑). Furthermore, Saturn is just past its morning rising, which further enhances the astrological portent.[10] The fact that Mars and Saturn are in positive, supportive aspects (Mars attends the Moon, and Saturn attends the Sun) is important in ensuring that the good aspects of either chart were not weakened by these pernicious planets. Either planet in quartile or opposition to the Sun or Moon could have undermined the conditions of that day.

Another time that produces incredible regal conditions for April 17, 6 B.C., is just before sunrise when the Moon is rising (a situation that, as we have seen, also occurs in Hadrian's horoscope). In his *Anthology,* Vettius Valens describes conditions that produce royal births, and his description is in excellent agreement with the astrological aspects of April 17, 6 B.C.

> When the Moon [is present in the Ascendant and] if the star of Kronos [Saturn] should also be present, he [the native] will lead many countries. If the star of Zeus [Jupiter] should also be present . . . , they will become great kings. (*Anthology* 2.4)

Valens also says that when the Moon and planets, particularly planets undergoing heliacal rising, are in the Midheaven, the conditions are regal.

> And if each of the stars [planets] should fall in this place [the Midheaven] when it is rising [heliacally] or when it has an application [conjunction or other aspect] of the Moon, sovereigns or kings are made, those who are leaders of countries, or they will be renowned in many places. (*Anthology* 2.7)

Valens confirms that a chart has great regal potential when there is a heliacally rising planet in conjunction with the Moon in the Midheaven. However, a configuration in which a heliacally rising

Jupiter is in conjunction with the Moon is of regal importance even when Jupiter is not overhead in the Midheaven.[11]

In any case, any horoscope drawn for April 17, 6 B.C., is regal because the Sun, the Moon, Jupiter, and Saturn are in the sign of Aries, conditions which perfectly and simultaneously fulfill the major regal principles! And amazingly, Jupiter, the Magi's star, was heliacally rising in the east and in a close conjunction (occultation) with the Moon, which gave even more regal significance to that day. Indeed, there is nothing ambiguous about the implications for Judea; the astrological conditions in Aries blatantly point to a regal birth there. Moreover, given that these extraordinary astrological conditions happened when people were hoping for freedom from tyranny and salvation from pagan intrusions, it is evident that the sky on April 17, 6 B.C., would have been thought to signal not just the birth of a Judean king but the anticipated birth of the Messiah.

A ROMAN ASTROLOGER'S CONFIRMATION

We might ask whether the regal conditions of April 17, 6 B.C., were rare or common. Rarity is one of those troubling concepts that have misled many investigators of the Magi's star. According to our modern expectations a special celestial event is one that happens infrequently. But, for several reasons, arguments about the statistical frequency of events cannot be used to evaluate their astrological significance. There are many possible sets of conditions that would produce a regal portent, and calculating the frequency of each set would be complicated. Moreover, some conditions have more regal significance than others. For example, imagine two horoscopes: one that holds three planetary rulers in a zodiacal trine and another that holds only two rulers in a trine but has an exalted planet. It is not easy to determine which horoscope would be more auspicious according to ancient beliefs without finding records interpreting and comparing the specific aspects that they contain. In fact, the subjective element of interpretation prevents any objective examination. There are many horoscopes with some degree of regal aspects. There was no sharp demarcation between regal and nonregal natal charts. As noted previously, some horoscopes with regal aspects may have been interpreted as signaling the birth of a magistrate, governor, or military leader. So statistical analyses are misleading and meaningless in Greek astrology.[12]

Historical records show that the seemingly unimpressive planetary alignments of April 17 were exactly what astrologers such as the Magi believed to constitute portents of a royal birth. To find the Star of Bethlehem we must avoid interjecting modern opinions by identifying "unusual" planetary alignments and instead determine conditions that the Magi would have deemed extraordinary. What did make a particular set of conditions uncommon was not its rarity but the number of important regal aspects that formed simultaneously and the absence of bad, destructive aspects.[13]

Even though statistical analyses cannot be applied, people may still wonder if the conditions that made up the Star of Bethlehem ever repeated. Although Jupiter is in the east in Aries every twelve years, the conjunction of Jupiter, Saturn, and the Sun happens only about every sixty years—once in a lifetime but not once in a millennium. Were any of these repeat occurrences as auspicious as the one in 6 B.C.? The next such conjunction happened on April 4, A.D. 54, when Jupiter was in the east in Aries. The horoscope for that day, illustrated in figure 24, also fulfilled many of the regal principles and conditions such as rulers of the trines and planetary attendance for the Sun. However, these seemingly great regal aspects were weakened by the appearance of Mars ♂ in Cancer ♋ in quartile aspect to the Sun, Jupiter, and Saturn in Aries ♈—an ominous threat. In fact, the prognosis is that a king born with this horoscope would lose his throne![14] Inspection of other similar conjunctions reveals either that one or more of the regal principles or conditions of April 17, 6 B.C., was missing or that the regal conditions were disrupted by a malefic planet, Mars or Saturn. All the evidence points to the incredible uniqueness of the Star of Bethlehem. So from an astrologer's perspective the Star of Bethlehem was indeed rare, but it was so on the basis of a set of complex rules that defy statistical quantification.

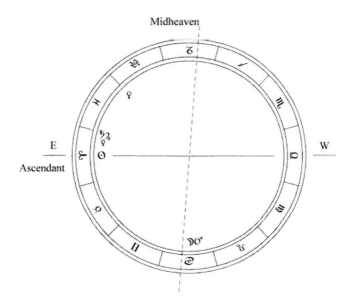

Figure 24. Mars ♂ maligned the regal portent of April 4, A.D. 54, by being in quartile to the Sun ☉, Jupiter ♃, and Saturn ♄.

Was the coincidence of so many regal conditions on April 17, 6 B.C., just fortuitous? As we have seen, I set out to find a special day on which occurred regal conditions that pointed undoubtedly to a magnificent birth under the sign of Aries. The lunar occultation of Jupiter served as my initial guess for an unusual astrological aspect. Perhaps my finding such an occultation on a day in the correct time frame was a coincidence. But once I found that Jupiter was rising in the east on that day, it became clear that I had uncovered a special day. Most important, the superb regal nature of horoscopes for April 17, agreeing perfectly with the hypothetical best case, pointed to cogent evidence that I had identified the correct day.

Although April 17 held the maximum astrological influences because on that day Jupiter rose in the east and was occulted by the Moon, some of the regal conditions persisted for several days or weeks. In fact, for several months other less important regal portents formed, portents that would have been noted by astrologers. Even the Bible mentions multiple portents. The star, namely, Jupiter, underwent retrograde motion ("went before") and became stationary ("stood over") in Aries on about December 19, 6 B.C. But the incredible conditions of April 17, 6 B.C., happened for sure only once during the reign of King Herod and the lifetime of Jesus.

In any case, searching for alternative regal days only repeats the mistake that modern investigators have made in looking for visually spectacular astronomical events. Trying to locate fascinating astrological alignments within the expected time frame still leaves open the possibility of finding only coincidences and interjecting modern expectations. Instead, confirmation that April 17, 6 B.C., revealed a Messianic birth must be found in the ancient astrological sources and not in speculative comparisons about the portentous regal nature of various days.

The many discussions in the astrological texts describing the individual astronomical events that together comprised the Star of Bethlehem have been discussed previously. However, these discussions consisted of generalized remarks about celestial configurations and alignments found in horoscopes of kings and emperors. What about specific mention of the horoscope of the Messiah? As it happens there is a reference in Firmicus's *Mathesis* to persons with a "divine and immortal" nature, a description that we would expect from people anticipating the advent of the Messiah.

Julius Firmicus Maternus, a Sicilian scholar, set out in ca. A.D. 334 to record in his *Mathesis* (Learning) the Greek astrological sources in Latin "for our Romans." In ca. A.D. 346, he wrote a second book, *On the Error of Profane Religion,* in which he attacked the pagan mystery religions from a Christian point of view. The *Mathesis* does not have any blatant references that would indicate whether Firmicus was a Christian; however, there appear to be references to the Jewish patriarch, Abraham, which may indicate that he was either familiar with Christian beliefs or in fact a Christian. Some historians think that he converted to Christianity between the writing of these two books. However, historian of astrology Jim Tester suspected that Firmicus was already a Christian when he compiled the *Mathesis*. At that time, after Emperor Constantine the Great had had a vision of the Christian cross at Milvian Bridge in A.D. 312, the Roman Empire was moving inexorably toward adopting Christianity. The slow replacement of paganism with Christianity culminated in the baptism of Constantine on his deathbed in A.D. 337. During this transitional period many people, including Constantine, continued to hold some of the old pagan beliefs and practices while professing their new Christian faith. Firmicus could have been a Christian even though he wrote unabashedly in the *Mathesis* about the influences of the stars of the pagan gods. Moreover, Tester saw the second book as Firmicus's reaffirmation of his Christian faith in response to the reception that the pagan references in the *Mathesis* probably received from pious Christians.[15]

It seems plausible that if Firmicus held Christian sympathies while writing the *Mathesis,* he may have made a reference to the Messianic portent that Christians saw as heralding the birth of Jesus. Even if he was not a Christian at the time, it is likely that he knew of that great day: he had access to a wealth of reference material, including horoscopes of Roman emperors. But in a style typical of other astrologers Firmicus rarely reveals the name of the person born under the astrological conditions he analyzes. Perhaps anonymity protected him from being accused of illegally monitoring the fate of an emperor or important person. Not mentioning names also allowed him to present the aspects as generalized examples that could be applied to other horoscopes.

There is a paragraph in the *Mathesis* describing several astrological aspects and conditions that pertain not just to regal conditions but to divine conditions. These center around Jupiter, and some can be related to the horoscope of Augustus Caesar, who was venerated as a god by Romans. However, at the end of the paragraph there are other astrological aspects that describe a different horoscope.

> If Jupiter comes into aspect with the waxing Moon, this will create men of almost divine and immortal nature. This happens when the Moon is moving toward Jupiter. It is difficult to observe this. If

Jupiter is in the north and the waxing and full Moon comes into aspect moving from the east (with Jupiter in his own house or exaltation or in signs in which he rejoices), the result is unconquerable generals who govern the whole world. This is especially true if the Sun in his exaltation is in trine aspect to Jupiter. For Jupiter rejoices by day when aspected by the Sun or Saturn, especially if he is in a morning rising. (*Mathesis* 4.3.9)

The condition described in the sentence about Jupiter being in the north (which means Jupiter was exalted in Cancer—a powerful condition) matches Augustus Caesar's horoscope. So does the condition about the waxing Moon (see Appendix D). Furthermore, Augustus was hailed as an unconquerable general governing the whole world, and his horoscope was well known during Roman times; thus, it is likely that Firmicus had that chart in mind. Relating this sentence to Augustus also makes sense because Augustus was declared divine by the Roman Senate after his death. However, the last divine portent, starting with the exalted Sun, does not fit his birth at all: Augustus was born when the Sun was on the border between Virgo and Libra, not when the Sun was exalted in Aries. Therefore, this other great portent must describe the birth of another divine and immortal person; and most significant, it matches the conditions that I have proposed for the Star of Bethlehem!

In fact, many of the conditions described in this paragraph by Firmicus are consistent with the conditions on April 17, 6 B.C., and this similarity strongly suggests that Firmicus had in mind the conditions of that portentous day.[16]

- "This [the creation of men of almost divine and immortal nature] happens when the Moon is moving toward Jupiter. It is difficult to observe this."—The movement of the Moon toward Jupiter is a process that takes a few days and, contrary to Firmicus's claim, is not difficult to observe. In fact, astrologers used calculations rather than observations to construct horoscopes. Their mathematics could easily predict an approaching conjunction weeks in advance. So why does Firmicus say "it is difficult to observe this"? That is, what could have made this observation difficult and what can be gained by watching a conjunction that was evident from calculations? I believe that Firmicus was calling attention to something that astrologers recognized as an important part of the conjunction, namely, how close the Moon comes to Jupiter. Mathematics could not predict the nearness of the encounter; this could be determined only visually.[17] But for an astrologer to observe this event, the Moon and Jupiter must be above the horizon on a clear night at the moment of closest passage, prerequisites making any visual verification difficult. Furthermore, the closest conjunctions, occultations, are very short-lived, lasting between an hour to less than a minute (for example, when a planet grazes the Moon's disk). Thus, I believe that Firmicus was calling attention not just to the motion of the Moon toward Jupiter but also to their ultimate separation, which may even produce the all-important occultation that modern mathematics verifies to have happened on April 17, 6 B.C. •

- "Jupiter . . . in signs in which he rejoices."—Jupiter is in the trine that it rules (Aries, Leo, and Sagittarius) and thus "rejoices" with extra astrological influences.
- "This is especially true if the Sun in his exaltation is in trine aspect to Jupiter."—The Sun was undoubtedly exalted in Aries—the only sign where it can be exalted. Furthermore, Jupiter was in the same trine and sign as the Sun. Moreover, they both ruled that trine.
- "For Jupiter rejoices by day when aspected by the Sun or Saturn, especially if he is in a morning rising."—No one knows whether Jesus was born during the day or during the night; however, Jupiter was indeed aspected by the Sun and Saturn (it was in conjunction with them). This sentence refers to the principle of attendant planets. And Jupiter and Saturn were protective attendants for the Sun. Finally, Jupiter had its heliacal rising on that day and thus began to be "in the east."

Thus, in this paragraph from the *Mathesis,* Firmicus describes the essential elements of April 17, 6 B.C. Most important, he claims that these marked the birth of not just a great ruler but a person with a divine and immortal nature. Furthermore, "unconquerable generals who govern the whole world" is a pagan metaphor alluding to a person with divine omnipotence. The term "unconquerable" did more than simply describe great military prowess (or luck); it had significant religious meaning for Romans who worshipped Sol Invictus, the Unconquerable Sun. Unconquerable generals, like Augustus Caesar, were believed to have supreme or more likely supernatural powers such as those of Sol Invictus. Similarly, to "govern the whole world" is another pagan metaphor. *Kosmokrator* (world ruler) was a sacred appellation reserved for the highest of divinities, such as Sol or Mithras.[18] Thus, this passage is about people with supernatural powers.

The religious significance of these allusions is illustrated on coins issued during Constantine's reign prior to Firmicus's writing

Figure 25. A bronze coin of Constantine the Great depicts Sol Invictus holding the cosmic orb as a world ruler (A.D. 317). Author's collection, RIC-7 134

of the *Mathesis*. The coin shown in figure 25 was struck in A.D. 317. This bronze coin is among the last Roman coins to display a pagan god.[19] Sol Invictus, the Unconquerable Sun, is shown with the inscription "Soli Invicto Comiti," which means that the Unconquerable Sun was a companion to the emperor. Below Sol is the mint mark of Trier, from whence the coin was issued. Most important, Sol is holding the cosmic orb (the universe, the "world"), which means that Sol is a *kosmokrator,* a world ruler.[20] If there had been a Roman messiah, he would have been referred to as an "unconquerable world ruler." Thus, this reference in the *Mathesis* has special relevance to the Star of Bethlehem and perhaps the birth of Jesus.

Because Firmicus converted to Christianity around the time that he wrote the *Mathesis,* it seems highly likely that he would have had Jesus in mind as the divine and immortal person to whom the conditions in the last part of the paragraph applied. That Firmicus may have placed Augustus Caesar and Jesus together among the divine may appear sacrilegious to Christians nowadays; however, this juxtaposition is consistent with the time period in which it occurred, during the transition of Roman beliefs from paganism to Christianity. That is, converted pagans saw relating two divine persons as logical and acceptable. But even if Firmicus was not leaning toward Christianity while writing this passage, it confirms that he was referring to records indicating that the horoscope of April 17, 6 B.C., would have been interpreted as signaling the birth of a divine, immortal, and omnipotent person. Thus, the excitement of the Magi recognizing this unusual portent is now understandable: "When they saw the star, they rejoiced with exceeding great joy."

ARIES AND THE ANTICHRIST

The coins of Antioch led me to the clue from the astrological sources that Judea was represented by Aries, the Ram, but by themselves the coins are meaningless. Their inscriptions give no hint as to why they were issued. Even the meaning of Aries is ambiguous according to the primary sources because that zodiacal sign may have represented either the entire Near East or Judea specifically. However, more Aries coins were issued during the reign of Emperor Nero. When they are analyzed in light of Greek astrology and historical records, these coins prove that Aries symbolized Judea and reveal why Nero was feared as the Antichrist.

In the fall of A.D. 54, forty years after Augustus Caesar died, Nero became emperor, and in that same year the governor of Syria, C. Ummidius Durmius Quadratus, started issuing large numbers of different coins in Antioch. On those coins Aries is seen for the first time since the death of Augustus.[21] On the old coins showing Aries, the god Jupiter had appeared on the obverse. But now, Tyche, the goddess of fate, graces the obverse, along with a Greek inscription that translates to "Of the people of the city of Antioch." On the reverse of the new coins Aries, the Ram, is in the same classic pose, leaping while looking backwards at a star. Underneath the leaping ram is the date, ΔP, which stands for "four and a hundred" years in the Caesarean era-system—A.D. 55–56 in our system. And the inscription around Aries can be translated roughly as "in the magistracy of Quadratus." During the same year a nearly identical coin was issued by Quadratus, but the star over Aries was replaced by a star and crescent (see figure 26).[22]

During Roman times, the star and crescent moon symbolized regal powers, namely, "majesty" or "sovereignty"[23] A star and crescent on coins did not represent an actual celestial event such as a lunar occultation of a bright star or a planet. Instead, it called attention to the powers of a king or ruler. On the Antiochene coins, the star and crescent symbol signifies a powerful ruler rising up under the sign of Aries. And that ruler, astrologers claimed, was the emperor Nero. Suetonius reports:

> Astrologers had predicted to Nero that he would one day be repudiated. . . .Some of them, however, had promised him the rule of the East, when he was cast off, a few expressly naming the sovereignty of Jerusalem, and several the restitution of all his former fortunes. (*Nero* 40)

Figure 26. A bronze coin of Quadratus issued in
A.D. *55–56 shows Aries with the star and crescent*
symbol, which represented majesty or sovereignty.
Author's collection, RPC-1 4287

If the rule of the East, or Judea in particular, was expressed as an astrological symbol, a star and crescent moon would be placed over Aries, exactly as depicted on the coins of Antioch.

The evidence that these coins from Antioch are not a coincidental curiosity is supported by coins issued in Damascus. The astrologers Ptolemy and Valens specified that Aries controlled, in particular, Coele Syria; and in A.D. 65–66 Damascus issued coins that were nearly identical to those from Antioch, that is, they bore Aries looking back at an overhead star and crescent moon.[24] This was the first time that Damascus made use of Aries with a star and crescent. Thus, during Nero's reign (A.D. 54–68) both Antioch and Damascus showed a special interest in the "sovereignty" or "majesty" over their zodiacal sign. This special attention for Aries on the coins and Suetonius's report of astrologers predicting Nero's return in the Near East can be explained by interpreting Nero's horoscope according to the principles of Greek astrology.

In Nero's horoscope (see figure 27) Mars ♂ rises immediately after the Sun ⊙, which is in the Ascendant (according to Suetonius,

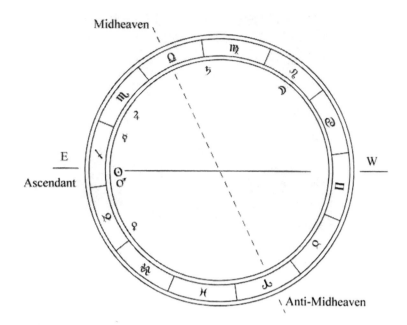

Figure 27. A reconstruction of Nero's horoscope for sunrise on December 15, A.D. 37, shows that Aries was in the anti-Midheaven, a position that indicated where his lost throne was to be recovered (Judea).

Nero was born at sunrise).[25] And Saturn ♄ is in quartile to them, which is an ominous configuration. Firmicus explains:

> If Saturn and Mars are in square aspect with Saturn above in the right square threatening Mars on the left [that is, if Saturn is three signs clockwise from Mars] . . . This combination also predicts loss of paternal inheritance. (*Mathesis* 6.9.4–5)

And Ptolemy corroborates the potential loss of possessions:

> But if the planets of the opposite sect overcome the governing places or rise after them, they bring about the loss of possessions. (*Tetrabiblos* 4.2)

Nero's chart fulfills this case for losing his possessions because Mars is a nocturnal *sect* (strongest influences at night), while the Sun is diurnal (strongest during day; see Appendix B).[26]

Nevertheless, two astrologers offer independent explanations that can be applied to Suetonius's report about the prediction of Nero's expulsion from Rome. Suetonius adds that Nero shrugged off this dire warning by making his famous remark, "a simple craft will keep a man from want." Although Nero may have been feigning humility—saying that his musical talents would sustain him after he was deposed—his faith in astrology assured him of a great comeback in the East.[27] And the reason that Nero would rise again to rule is also interpretable from the astrological records, which show that astrologers looked to the sign in the anti-Midheaven (Imum Caelum) to foretell where he could recover his losses. Firmicus says,

> This house [anti-Midheaven] shows us family property, substance, possessions, household goods, anything that pertains to hidden and recovered wealth. (*Mathesis* 2.19.5)

Dorotheus of Sidon elaborates on this idea:

> There were some of the ancient scientists who looked concerning the matter of theft from the four cardinal points, and if one of them was asked about a theft or something lost he would look concerning what was stolen or lost from the Ascendant [rising sign] . . . and where they put the goods from the cardinal point under the earth [anti-Midheaven]. (*Pentateuch* 5.35.20)

Thus, astrologers scrutinizing Nero's horoscope found the Sun in the Ascendant, which they interpreted as representing his paternal inheritance, namely, the imperial throne.[28] But the rising Sun was "under attack" by Mars and Saturn. Hence, the astrologers looked to the anti-Midheaven to find out where Nero's lost throne would be recovered, and there they found Aries (see figure 27) and thus specified the Near East for his return. Most important,

Figure 28. The city of Ptolemais issued this bronze coin in ca. A.D. *66–68 honoring Nero's astrological majesty by displaying the star and crescent with him. Author's collection, RPC-1 4750*

Suetonius's account shows that some astrologers believed that Judea was represented by the same sign that others recognized as representing the East: Aries, the Ram.

With this information in hand, we can now interpret the coins of Antioch and Damascus bearing Aries with a star and crescent. The theory is that the coins signify the prediction that Nero would rise up as a monarch in the Near East. That is, the star and crescent symbol conveys the regal power of majesty, and Aries symbolizes the Near East. That Nero was associated with the majesty bestowed by the regal astrological symbol is evident in figure 28, which shows a coin from Ptolemais (see map 1). On the coin the star and crescent symbol is shown before Nero, who is wearing the wreath of laurel leaves, a victorious insignia; and the inscription hails Nero as *Imperator,* "commander." This coin verifies for us that the astrological symbol of majesty or sovereignty was related particularly to Nero. Thus, the coins of Antioch and Damascus bearing Aries and the star and crescent allude to Nero's prophesied reign in the Near East or, as some astrologers said, Jerusalem. The assertion by astrologers that Nero would rise up in Jerusalem would produce serious repercussions for some people.

Nero never made it to the East: he committed suicide in A.D. 68 near Rome. Nevertheless, many people must have believed the astrological prediction: it produced several important stories that come down in various historical records and even in a biblical account. For instance, immediately after Nero's death, impersonators sprang up in the eastern Roman Empire. One was an accomplished musician, like Nero, and even resembled the emperor. Thus, some people—such as the Greeks, who knew Nero as a great friend of their culture and country—rallied to support the impersonator. But he was killed, as were the other charlatans.[29]

The prophecy of Nero's return was also noted by Jewish and Christian seers who warned about the return of the feared and hated emperor. In the *Sibylline Oracles,* a collection of prophecies compiled by Alexandrian Jews and Christians, stories about the Messiah and anti-Messiah (Christ and Antichrist) were told, and Nero was cast in the role of the evil nemesis of the Messiah.[30] Many Jews blamed Nero for the devastating revolt caused by his mishandling of Jewish relations and the appointment of notorious procurators in Judea. And Christians remembered him for blaming them for the fire that destroyed Rome, blame that gave rise to the Great Persecution. The fourth book of the *Sibylline Oracles,* written in A.D. 80, reports that Nero did not die but rather fled across the Euphrates River in Mesopotamia; and the book ominously predicts that he will return and rule again. The possibility of Nero's return to Judea was truly frightening for Jews and Christians.[31]

Biblical researchers note that Nero was cast in the role of the Antichrist in The Revelation of Saint John the Divine. An unknown author from the end of the first century A.D. writes about the beast whose number is 666, which is most likely Nero because these Hebrew numerals translate to Nero's name. Furthermore, this biblical account claims that the beast "had the wound by a sword, and did live," which leads many people to believe that Nero is the subject of the prophecy about the Antichrist of the

Apocalypse.[32] Some scholars claim that Nero's astrologer, Balbillus, was also vilified in Revelation. Pious Christians saw the astrologer as Nero's chief advocate. Thus, Balbillus was denounced because "he had the power to give life unto the image of the beast."[33] That is, his astrological prognostications propagandized Nero's glorious recovery in Jerusalem.

It is apparent that Greek astrology has a connection to two important prophecies. The Star of Bethlehem was an astrological portent of extraordinary significance, one that would have been noted by Christians as the Messianic star fulfilling Balaam's prophecy. And Greek astrology is the source for the story about Nero recouping his lost throne in Jerusalem, which was interpreted by Christians as the advent of the Antichrist. Most interestingly of all, both stories are tied together by Aries, the Ram, which symbolized not just the Near East but specifically and more significantly Judea.

CHRISTIANS IN ANTIOCH

It is not easy for us to change our views about the Star of Bethlehem, which have been shaped by stories that popularized and mythicized the star and by strong religious views about Jesus. Nevertheless, my new findings about the star have important historical and religious implications.

The evidence shows that astrologers did indeed recognize the Star of Bethlehem as an extraordinary celestial portent of a royal and divine birth. In particular, the information gleaned from Greek astrology, the kind of astrology practiced during Roman times, reveals elements in the account in Matthew that point to a historical basis that is stronger than previously believed. The literal translations of the terms describing the Magi's star obscured the underlying astrological meanings that are recognizable in the older, Greek text of the Bible. As we have seen, one of those terms, "in the east," was cited twice in Matthew because it was the most important astrological phase for the regal planet, Jupiter. Jupiter's appearance as a morning star occurred on April 17, 6 B.C., and was accompanied by other powerful astrological aspects that pointed to a magnificent star-blessed birth.

The seemingly mysterious phrases "went before" and "stood over" were also Greek astrological terms that defined secondary, but less important regal conditions, which may have reinforced the previous regal condition, that is, Jupiter's heliacal rising. Such technical terminology in Matthew shows that the evangelist had access to sources well versed in astrology. But above all, these details give further credence to the historicity of the story. Not only did Jupiter undergo all of these conditions in the expected time frame of Jesus' birth, but also there were other strong regal conditions that focused attention on Aries, the Ram—the sign of the Jews.

These technical matters in Matthew also shed some light on the controversial story about the Slaughter of the Innocents—Herod's alleged killing of children who were born under the regal influences of the Star of Bethlehem. As we have seen, it is puzzling that children two years and younger were seen as threats to Herod. But the answer to this mystery lies in the lesser regal portent to which the Bible referred, the portent that happened eight months after Jupiter was in the east. Jupiter's retrograde motion and subsequent stationing on December 19, 6 B.C., can be thought to be relatively superfluous compared to the truly auspicious time when Jupiter was in the east, on April 17, 6 B.C. But now their significance becomes apparent: Herod was threatened by children born on more than one regal day. Astrologers would have explained that although children born under the lesser regal portent may not have been the Messiah, they were nonetheless a danger to Herod's throne.[34] So, the evangelist tells us, Herod struck out against children born under different regal conditions. There may have been even more potentially regal days during that time, days that fed Herod's well-known paranoia.

These findings also show that there is no reason to question the date of Herod's death, as some have done. The vast majority of historians claim that Herod died in late March or early April of 4 B.C., two years after the emergence of the Star of Bethlehem on

April 17, 6 B.C. The children born under regal stars were two years old and younger, which indicates that Herod was alive for at least two years after the star's first manifestation. Some people may now wonder whether his timely death so close to two years after the star appeared in the east interfered with the executions of these two-year-old children and may explain why the murders cannot be verified by historical records. But that is conjecture. Although the executions cannot be corroborated by other sources, there is now more information that clarifies some elements of the story, information that makes a stronger case for a historical basis for those elements.

People have also wondered why the account in Matthew referred to astrologers: astrology lies beyond customary Judeo-Christian beliefs and mores. In fact, Greek astrology had many pagan elements, so its discussion in the Bible seems inappropriate.[35] The response that biblical scholars readily give is that astrology had no role in the Star of Bethlehem, that instead the star fulfilled the prophecy of Balaam—"a Star out of Jacob"—a prophetic star that Jews anticipated. That may be so, but the role of astrology was important as well. That is, the message of Matthew, which contained astrological terminology, was meant not just for Jews but also for the larger Hellenistic world, where astrologers wielded powerful influence and astrological prognostications would have been heeded unhesitatingly. The biblical account shows not only that the star fulfilled the Jewish prophecy of Balaam but also that the birth of the King of the Jews was validated by non-Jews practicing a respected art, namely, astrology. Had we remembered the importance of astrology in the Hellenistic world, we would not have questioned its discussion in Matthew.[36]

Although no source has survived the ages to allow us to unambiguously determine when Jesus was born, the evidence I have presented points to April 17, 6 B.C., which produces a believable chronology of Jesus' life. This birth date would make Jesus two years old when Herod died. He would have been thirty-three or thirty-four years old when he started his ministry in the fifteenth year of the reign of Emperor Tiberius (A.D. 28–29), which agrees with Luke 3:23 that Jesus "was about thirty years old" at that time. And Jesus would have been thirty-five to thirty-eight when he was crucified in A.D. 30–33. This chronology agrees reasonably well with the biblical evidence.

Some may still argue that although there was a Star of Bethlehem, there is no incontrovertible proof that Jesus was actually born on the day that it appeared. The evidence from Firmicus is truly enticing, but we cannot say conclusively that he was describing Jesus' birth. Early Christians and astrologers such as Firmicus may have believed that Jesus was born then because the star was a Messianic portent, but there is no independent record corroborating Matthew that Jesus was born on the day of the remarkable portent. There may have been astrologers searching for the Messiah in Judea, but we cannot confirm that they indeed found Jesus and verified that he had been born under this portent. Some people may argue that Jesus may have been born close to the time of the star and that pious Christians thus presumed that he was born then. Thus, the story could still be a midrash interpretation in which two unrelated events (the Messianic star and Jesus' birth) were tied together years later in Matthew to strengthen the message that Jesus was the Messiah.

Of course, my findings cannot validate all of the details in the Bible about the birth of Jesus. For instance, the astrological evidence cannot prove that Jesus was born in Bethlehem: anyone born anywhere on April 17, 6 B.C., would have had a regal birth. It is precisely because the Magi were not sure exactly where the birth had occurred that they went to Jerusalem, the capital city of the future king, to ask Herod for advice about the likely birthplace. However, many biblical scholars feel that Bethlehem was cited only because the Messiah had been prophesied to come from the lineage of King David, whose family town was Bethlehem.

There is, however, another part of the infancy narratives that my findings do help to clarify: the account of Luke, which has puzzled biblical scholars even more than the Matthean account has. The evidence according to those scholars is that the birth of

Jesus was not during the census of Quirinius as Luke places it, but during King Herod's reign. As noted previously, other evidence in Luke also places Jesus' birth in the reign of Herod. An intriguing answer to this dilemma about Quirinius comes from the city of Antioch.

Antioch was truly the cradle of the Christian Church. There, Christianity separated from Judaism, and the followers of Jesus were first called Christians.[37] After the crucifixion of Jesus (A.D. 30–33) his disciples began preaching that he was Christ, the Messiah. But their conversion of Jews met stiff opposition from many traditional Jews, who saw the new principles of Jesus as a threat to the fundamentals of Judaism. The arguments between the traditional Jews and the followers of Jesus became violent, and Stephen, a disciple of Jesus, was stoned to death by those who were opposed to these new principles. This incident was followed by a persecution of Christian Jews, who then fled Jerusalem and gathered in Antioch.[38] The Church in Antioch began to convert non-Jews, and Paul the Apostle used Antioch as a base for his missionary work. Biblical scholars believe that the stories about Jesus' birth from Matthew and Luke were written (A.D. 80–90) in the growing Christian community of Antioch. And it is the influence of astrology in Antioch that presents an explanation for the Lucan account.

Perhaps when the Romans of Antioch first placed Aries upon their coins (ca. A.D. 5–11), their only intention was to proclaim the glory of Antioch, or even to celebrate the annexation of Judea in A.D. 6. But it is puzzling that they used the zodiacal sign of Aries when only a dozen years earlier, in 6 B.C., Aries was the site of the Star of Bethlehem. And given that astrology was important in the Hellenistic city of Antioch, it incomprehensible that the Romans who issued these coins did not have the Star of Bethlehem in mind, given the incredible significance it held. That is, because astrology was an integral part of their culture and beliefs, the people of Antioch had to have been aware of what had happened in Aries in 6 B.C. Of course, no Roman would strike coins commemorating the birth of the King of the Jews. Coinage was usually exploited as propaganda that advanced moral or religious themes, local pride, or important political events. A plausible theory is that the Romans were indeed sensitive to rumors about the Messiah born under a regal star in Aries and that the coins may be in fact a record of their reaction: to present an alternative interpretation of the portent, to say that the heavens foretold the annexation of Judea, not the advent of the Messiah. Another plausible idea is that the annexation of Judea and Samaria meant that all astrological manifestations in Aries now focused only on Antioch (or Roman Syria). That is, the annexation ensured that any portent in Aries, the Ram, was meant exclusively for the Romans of Antioch, not for the Jews. Thus, Aries on the coins alludes to the new destiny of the expanded province and asserts Antioch's ownership of Aries. This would have been an important claim in a society that embraced astrology and was wary of rumors about the Messiah rising up under Aries.

Even if these theories about Roman intentions in using Aries are not entirely correct, Christians in Antioch examining the coins would have nevertheless thought that they depicted the Star of Bethlehem. That is, converted pagans knowledgeable about astrology would have remembered stories about the Messianic portent, but they probably would have been uncertain about exactly when it happened. The plausibility of this theory leads to an explanation of the census of Quirinius in the Lucan account. First, as just noted, the coins display the zodiacal sign that was known to be where the Star of Bethlehem had manifested itself. That is, astrologers in Antioch and throughout the Near East would have told people about this remarkable portent in Aries and therefore people would naturally have recalled the Star of Bethlehem when looking at the coins. Second, there is the uncanny similarity between the scene in Luke of "shepherds abiding in the field, keeping watch over their flock by night" and the coins depicting Aries, the Ram, under a star. A person unfamiliar with astrology might think that the coins show a sheep under a starry night sky. Or a person who did not believe in astrology might have intentionally given the coins' scene a non-astrological interpretation, again a sheep at night rather than Aries under a fateful star of a pagan deity.[39] And third, biblical scholars tell us that the term "heavenly host" used in Luke is a reference to stars and planets, specifically the morning stars

> And suddenly there was with the angel a multitude of the heavenly host praising God, and saying, Glory to God in the highest, and on earth peace, good will toward men. (Luke 2:13–14)

That is, the evangelist of Luke, like that of Matthew, was telling a story about a celestial Messianic portent involving the morning stars (Jupiter and Saturn) but did not make any specific references to astrologers as the Matthean account did. Instead, he wrote about the planets as messengers of God, namely, the heavenly host announcing the birth of Jesus as the Messiah.[40]

Believing that the coins portrayed the Messianic star under which Jesus had been born, the people of Antioch would have spread stories and perhaps the evangelist of Luke used those stories to reconstruct the Nativity. Thus, the evangelist placed the time of the birth according to when the coins with Aries first appeared, around the time of Quirinius's governorship. This explanation is more plausible than the theory that the evangelist confused the tumultuous times of Herod's death with the upheavals surrounding Quirinius's annexation and taxation.

My research on the Magi and their special star reveals much more agreement between Matthew, historical records, and astrological sources than previous inquiries have, agreement that fortifies the case for a historical basis for the Star of Bethlehem in the account in Matthew. Not only has the level of verisimilitude, the likelihood of the story's truth, been raised but also nothing has been revealed to undermine the validity of the story. Even the Lucan account is more understandable if we assume that the evangelist was trying to reconstruct the portent signaling the birth of Jesus. Luke describes the portent as a gathering of the heavenly host, a description that avoids any references to pagan astrology.

Furthermore, new significance can now be given to records that were previously given little weight and new avenues of investigation may open. There is, for instance, the intriguing date of Jesus' birthday cited by Clement of Alexandria, namely, April 20 or 21. We do not know to which year Clement was referring, but in 6 B.C. both of these dates were portentous, but slightly less so than April 17.[41] If the birth occurred just three or four days after Jupiter's heliacal rising, some of the significant aspects of the portent would still have been in place.

Because no one could identify a believable celestial event, the account of the Magi's star has up to now been challenged as a midrash convention, a fable, or a pious dream. And some truly incredible ideas have been advanced, ideas that reinforced opinions that the account was a myth. Undoubtedly, the new evidence that I have uncovered will have important consequences for Christians and for scholars. People can be assured that the Star of Bethlehem did indeed exist, which may even strengthen the faith of some. For others this work may lend greater understanding of history, ancient practices and culture, and the origins of beliefs about the sky. I hope that it will give new direction to the study of ancient astrological records now that their usefulness has been demonstrated by my investigation of the Star of Bethlehem.

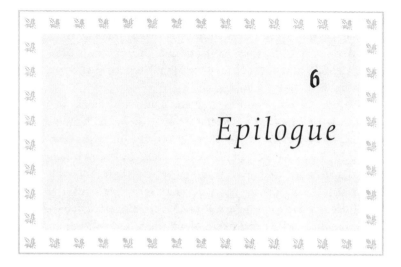

6

Epilogue

Having shared my findings, I close with a personal note. For me the story about deciphering the Star of Bethlehem began, and now ends, with the coins of Antioch. When I first examined these coins, I had no idea that they would lead me to investigate the Magi's star. I will never forget my astonishment when I first consulted Ptolemy's *Tetrabiblos* for the meaning of Aries on one of my Antiochene coins. Believing that researchers had established that Pisces, not Aries, represented Judea, I was troubled by Ptolemy's indication that the coin's symbolism was about Judea rather than Antioch; after all, the coin came from Antioch, not Judea. After much reflection, I realized that I was forcing incorrect assumptions onto the coin's symbolism, that I had to accept what an astrologer from two millennia ago was telling me: Aries had been the sign of the Jews. That acceptance pointed to the coin's connection to the annexation of Judea by the Romans of Antioch. After some further thought I realized that the clue to the biblical star was sitting in my coin box: the theories about the Magi's star having appeared in Pisces were wrong.

My experience with the Aries coins forced me to study the ancient astrological records, which was a challenging task for this astronomer who does not believe in astrology. Perhaps my skepticism gave me a fresh point of view; because I refused to accept someone else's opinion about ancient beliefs, I felt compelled to verify everything from primary sources. In the course of this research I found that astrology had been an integral part of our ancient culture. The astrologers of antiquity were the scientists of their time, trying to make sense out of the world around them. For the most part, they were indeed "wise men" who earnestly set out to devise a system to guide them and help them to understand their world. Moreover, in their role as imperial and royal advisors, they greatly shaped historical events. My studies have convinced me that the influence of astrology in Western history and culture has been greatly ignored (perhaps intentionally) and that this lack of attention has only obfuscated the truth. And I see no reason why we should hide or ignore the historical importance of astrology. Only by being open-minded about astrologers' work can we find the truth behind our history and heritage. Their almost forgotten legacy becomes apparent in the stories about the birth of Jesus: astrology not only contributed to the fashioning of the Matthean account about the Magi and their star but also is behind the Lucan account about the heavenly host and the "shepherds abiding in the field, keeping watch over their flock by night."

The Star of Bethlehem turned out to be very different from what I expected. Like other people, I had anticipated something visually dramatic rather than arcane and cerebral. Nevertheless, we can be assured that the extraordinary conditions of April 17, 6 B.C., were as real and dramatic as any blazing comet or exploding supernova. Unlike those spectacular but terribly foreboding or meaningless apparitions, the portent formed in Aries by Jupiter and other celestial bodies conveyed a joyous, wondrous message about a regal and divine birth in the kingdom of Herod the Great. Now we can look at the star above a crèche at Christmas and know that there was indeed a Star of Bethlehem.

Appendices

APPENDIX A:
DEFINING THE POSITION OF THE ZODIAC

Although the earth's orbit about the Sun has remained fairly constant over the millennia, the orientation of the earth with respect to the plane of its orbit has perceptibly changed. The rotational axis of the earth precesses; that is, the spinning earth slowly wobbles like a toy top. As the axis of the earth precesses, it points to different stars; and thus the pole star changes every few centuries. This precession shifts the intersection of the celestial equator and the ecliptic and moves the seasons through the constellations along the zodiac. This drift of the seasons, called the precession of the equinoxes, amounts to a shift the size of the Sun's disk (one-half degree) in about thirty-six years. It takes about twenty-six thousand years for the vernal equinox to return to the same position on the ecliptic.

Hipparchus is credited with recognizing in ca. 130 B.C. that errors in stellar positions were due to the precession of the equinoxes. His great star atlas cataloguing the brightness and position of the so-called fixed stars (an atlas that was inspired by a nova in 134 B.C.) was finished in 128 B.C. and revealed that the entire celestial sphere had changed its course slightly over the centuries. This finding led Hipparchus to discover the precession of the equinoxes. Although this is a tiny effect, it produces discrepancies that grow over the centuries. It is customary to measure celestial positions relative to the vernal equinox, and if the position of this benchmark is not carefully determined, the accuracy of celestial positions is lost over time. During late Babylonian times, the spring equinox was located within Aries, the Ram, ranging from the eighth to the tenth degree over the centuries. And Roman and

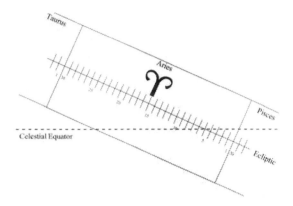

Figure 29. At the beginning of the Christian Era astrologers placed the vernal equinox near the fifth degree of the sign of Aries ♈.

Greek astrologers adopted this convention in principle, although they made adjustments for the drift over the centuries. But the data from horoscopes indicate that the offset had been reduced to the fifth degree by the start of the Christian Era. However, it was Ptolemy in ca. A.D. 150 who permanently fixed the beginning of Aries with the vernal equinox and let the zodiac drift with the precession of the equinoxes. Today, this tropical zodiac, defined by the vernal equinox, no longer coincides with the physical zodiacal constellations. In keeping with the procedure established by Ptolemy, modern astrologers use a tropical zodiac to cast their horoscopes. In this analysis of the Star of Bethlehem, ♈ 5° = v.e. was adopted according to Neugebauer's findings (see figure 29). My own analysis of horoscopes from this time indicates that his offset of the vernal equinox is the upper limit, with a lower limit of ♈ 3° = v.e. Nevertheless, the conclusions are valid for ♈ 2° ≤ v.e.♈ ≤ 6°, which is a comfortable margin of error.[1]

A P P E N D I X   B :   A S T R O L O G I C A L   I N F L U E N C E S

The astrological influence of a celestial body was determined, in part, by its aspect, that is, its geometric relationship, with other celestial bodies. Although the *trine* was the most important of the aspects, there were other significant geometrical configurations. The *sextile* (hexagon composed of signs separated by sixty degrees) was important, but its influence was weaker than that of a trine. The trine and sextile were positive influences, but the *quartile* (square made by signs separated by ninety degrees) was potentially negative, especially if Mars or Saturn was in quartile with the Moon or Sun or if they were in quartile with each other. Planets in signs diametrically across from each other in the chart were said to be in *opposition*. The influence of this aspect was determined by means of a complicated set of rules, rules which could sometimes mitigate a potentially bad condition. The *conjunction* (the appearance of celestial bodies in the same sign) was understood to be a powerful and important positive alignment, but Ptolemy and other ancient astrologers did not consider it to be an aspect per se, probably because it was not a geometric pattern such as a square or triangle.[2] Moreover, the conjunction is effectively a special case of a trine: planets in the same sign are also in the same trine.

Another relationship governing the influence of a celestial body was its position relative to its *domicile,* or zodiacal *house.* The two luminaries and the five planets were assigned signs in which they ruled, and when a planet was in its sign its influence increased (a planet in its domicile was not, however, as important as a planet in its exaltation). Each of the five planets ruled over two signs, one by day, the other by night. This system has been revamped by modern astrologers to account for Uranus, Neptune, and Pluto, planets that were unknown in ancient times. The original Greek scheme of the domicile rulers is shown in figure 30. From the figure we can see that Mars ♂ rules in Scorpio ♏ during the day and Aries ♈ during the night.

Each zodiacal sign is also subdivided into zones in which a celestial body has special power. Moreover, there are several systems of division (decans, monomoiria, etc.), but it is the system of *terms* that was prevalent in Greek astrology. Each of the five planets rules over small sectors, called terms, in which it holds extra influence. As with many other astrological systems, there are

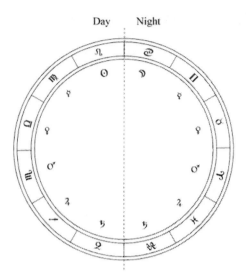

Figure 30. The domiciles were defined according to this scheme, which has been altered over the millennia to accommodate new planets.

several different schemes. The most popular one was the system of Egyptian terms. Another system was the system of Chaldean terms. Vettius Valens and Ptolemy had their own systems as well. In the analysis of the Star of Bethlehem the terms were not considered: they played a minor role in regal horoscopes, and it is not clear which system of terms should be applied. In addition, as we have seen, the uncertainty about the position of the vernal equinox complicates the definition of the boundaries of the sign and, therefore, of the small terms as well.

The celestial bodies were said to possess special powers over certain areas of human concern. Table A.1 lists the significant characteristics assigned by Ptolemy and accepted by most other astrologers.[3] These astrological influences can be magnified or annulled by interactions between celestial bodies and between celestial bodies and the zodiacal signs.

Three additional properties of luminaries and planets were sig-

Table A.1. General Astrological Characteristics of the Celestial Bodies

BODY AND SYMBOL	GENERAL PROPERTY
Sun ☉	Honor, justice, success, dignity, reverence
Moon ☽	Character of the soul, natural endowments, activity
Mercury ☿	Religious matters, changes in laws and customs, shrewdness, ingenuity, robbery, piracy
Venus ♀	Abundance, marriage, children, order
Mars ♂	Wars, violence, arson, enslavement
Jupiter ♃	Fame, prosperity, happiness, political power, greatness
Saturn ♄	Illness, fear, deaths, famine, destruction

Table A.2. Nature, Sect, and Gender of the Luminaries and Planets

NATURE		SECT		GENDER	
Beneficent	♃ ♀ ☽	Diurnal	☉ ♃ ♄	Masculine	☉ ♃ ♄ ♂
Maleficent	♄ ♂	Nocturnal	☽ ♀ ♂	Feminine	☽ ♀
Variable	☿ ☉	Both	☿	Either	☿

nificant: *nature, sect, and gender* (see table A.2). Jupiter and Venus are good-natured, Saturn and Mars ill-natured, but the latter can produce good results under certain conditions. Mercury lives up to its name; the planet is truly mercurial in all categories. Ptolemy adds the Moon to the group of beneficent celestial bodies, and he tells us that the Sun also has a variable temperament. However, most other astrologers did not classify the Sun or Moon. A planet's sect—diurnal, nocturnal, or both—has bearing on its influence. We are told that the Sun, Jupiter, and Saturn "rejoice" with extra influence during the day; they are said to be diurnal. The Moon, Venus, and Mars are nocturnal; thus, they play important roles for the nighttime. The zodiacal signs were also assigned sect and gender (see table A.3), and astrologers believed that good conditions arose when the gender or sect, or both, of a celestial body matched that of the zodiacal sign in which it was

Table A.3. Gender and Sect of the Zodiacal Signs

Gender and Sect	Sign
Masculine and diurnal:	♈ ♊ ♌ ♎ ♐ ♒
Feminine and nocturnal:	♉ ♋ ♍ ♏ ♑ ♓

located. For example, masculine and diurnal celestial bodies often do best in masculine and diurnal signs.

A P P E N D I X   C :
T H E   S Y S T E M   O F   M U N D A N E   H O U S E S
A N D   C A R D I N A L   P O I N T S

The important features of the system of *mundane houses* and *cardinal points* are the *angles, centers, cardines,* or as used in this book, *cardinal points.* These consist of four important zones in the sky: the rising sign on the eastern horizon (the Ascendant), the highest zodiacal sign (Midheaven, *Medium Caelum,* or upper culmination), the setting sign on the western horizon (Descendant, *Dusis,* or setting), and the lowest zodiacal sign (anti-Midheaven, *Imum Caelum,* subheaven). These terms apply not only to these zones but also to the lines that divide the zones. The cardinal points most likely stem from an earlier, simpler form of astrology that did not have the capability to differentiate positions finely. Most important, the four cardinal points serve as the central axes that divide the zodiacal circle into a system of *loci,* or *places.* Modern astrologers refer to this as the system of houses, which is sometimes confused with the system of domicile houses. Thus, the *loci* will be called mundane houses to distinguish them from the domiciles. As it turns out, only the all-important cardinal points are needed for investigating regal horoscopes (although the other mundane houses can also support regal conditions).

The effects and degrees of influence of the mundane houses are listed in table A.4. There are seven good, beneficent houses and five evil, maleficent houses. It is possible to rank each house's influence such that +1 is for the best of the beneficent houses, which turns out to be the first house, and +7 for the weakest, the ninth house. Likewise, the most maleficent, the twelfth house, is −1,

Table A.4. The System of Mundane Houses

Number	Name	Rank of influence	Influences
1	*Ascendant* "Life"	+1	Character of the birth; best beneficent house; primary cardine
2	"Hope" "Gate of Hell"	−4	Personal hopes and material possessions, livelihood
3	"Dea" and "Brothers"	−5	Brothers, friends, and travel
4	*Anti-Midheaven* "Parents"	+6	Family property, recovered possessions, retrieved wealth
5	"Children" and "Good Fortune"	+4	Number and sex of children
6	"Health" and "Bad Fortune"	−2	Physical infirmities and sickness; almost as evil as twelfth house
7	*Descendant* "Spouse"	+5	Nature and number of marriages
8	"Death"	−3	Material power and worldly position; kind of death
9	"Deus"	+7	Social class, religion, and foreign travel
10	*Midheaven*	+2	Professional careers, life and vital spirit, all actions, country, home, dealings with others; almost as powerful as the first house
11	"Good Daemon"	+3	Accomplishments; good sextile aspect to Ascendant
12	"Bad Daemon"	−1	Nature of enemies, character of slaves, defects, illness; most evil of houses

while the weakest, the third house, is −5. This ranking system is adapted from Dorotheus of Sidon and Hephaestion of Thebes,[4] although it agrees with the rankings of most other astrologers. Interestingly, there are more beneficent houses than there are maleficent, which apparently reflects an optimistic view of Fate. In any case, the two most important beneficent houses are the first and tenth houses—the Ascendant and Midheaven, respectively. These houses figured prominently in regal horoscopes.

There are slight differences of opinion on the exact positions of the cardinal points. The Ascendant customarily starts at the rising point (Horoscopus) and runs thirty degrees below the horizon. A vertical axis runs from the location of the newborn person up to the Midheaven and extends down to the anti-Midheaven. (The astrological abbreviations for the important houses and cardinal points are ASC [Ascendant], MC [*Medium Caelum*, Midheaven], DSC [Descendant], and IC [*Imum Caelum*, anti-Midheaven].) Ptolemy slightly skewed his cardinal points by five degrees so that the Ascendant started five degrees above the horizon and dipped to twenty-five degrees below.[5] This skewing ensured that the rising point, the Horoscopus, lay within the Ascendant cardine.

In some horoscopes, the Midheaven line is shown not perpendicular to the horizon but tilted, depending upon the observer's latitude. If an observer moves north or south, the Midheaven in the sky remains fixed, but the zodiacal point on the Ascendant will shift, changing the angle between the Ascendant and the Midheaven and tilting the Midheaven line in a horoscope. The same process causes people living in the northern latitudes to see sunrise earlier during the summer than those who live farther south but still at the same longitude (and thus the more northerly people experience a longer day). In this case the angle between the Horoscopus (where the rising Sun is) and Midheaven (where the Sun will be at noontime) is larger for people who live farther north.

APPENDIX D: REGAL HOROSCOPES

Augustus Caesar

The birth date of Augustus Caesar has been the subject of much scrutiny and controversy; although it is generally cited as just before sunrise on September 23, 63 B.C., on the Julian calendar (which was, as we have seen, introduced by Julius Caesar in 46 B.C., after the birth of Augustus).[6] Although Augustus wrote that his birthday was September 23, researchers have shown that September 22, in particular, produces a regal chart.[7]

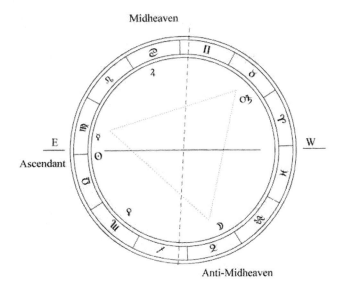

Figure 31. A reconstruction of Augustus Caesar's natal
horoscope shows that the Moon, one of the night rulers of
Trine II—Taurus ♉, Virgo ♍, and Capricorn ♑ was in
Capricorn.

The generally accepted horoscope is shown in figure 31.[8] The
Midheaven line is tilted because of the geometric effects produced by projecting the celestial sphere onto a plane (see Appendix C). At first glance, Augustus's astrological chart appears
ordinary; it contains no spectacular astronomical events such as
the triple planetary conjunction that some have proposed as the
Magi's star. And not surprisingly, there is no record pointing to
either a comet or a nova blessing his birth: these did not figure
into regal births. Nevertheless, we must realize that Augustus's
chart ranks as one of the greatest in history and is the epitome of
the regal horoscope.[9] From an astrologer's perspective there are
several indisputably important features. Jupiter ♃ is exalted in
Cancer ♋. Firmicus in his *Mathesis* tells us that when Jupiter is in
its exaltation a regal condition exists:

> He will make the native high-born, famous, always ruling great
> states, perhaps the first ten of great states. (*Mathesis* 3.3.1)

In Augustus's horoscope, the Sun ☉ is in the Ascendant, the most important cardinal point.[10] Because Augustus was born at night, that is, before sunrise, the Moon rules the birth. In the horoscope, the Moon is in Capricorn, which is part of the trine of Taurus, Virgo, and Capricorn (♉, ♍, ♑; see dotted line in figure).

So the following conditions for the regal principles emerge from Augustus's chart:

- Exaltations: Jupiter is in its exaltation in Cancer. Mercury is exalted in Virgo.
- Trines: The Moon is in Capricorn, which is in the trine of Taurus, Virgo, and Capricorn. The Moon rules this trine at night, and Mars is a co-ruler. That Mars is also in its trine is another great condition. Nevertheless, it is the presence of the ruling Moon in Capricorn that is indicative of Augustus's regal birth.[11]
- Cardinal points: The Sun is in the Ascendant; the Moon is at the anti-Midheaven; and Jupiter is in the Midheaven. The presence of these three important bodies in the cardinal points is a powerful portent.
- Attendants: Mars protects the Moon by setting after it, and the two exalted planets, Mercury and Jupiter, rise before the Sun, conditions that fulfill the requirements for attendance.[12]
- Jupiter: The regal planet's prominent role in bestowing the emperorship upon Augustus is strengthened by the fact that it attends the Sun. Jupiter was exalted and in the all-important Midheaven. Finally, Jupiter is in the trine Cancer, Scorpius, and Pisces (♋, ♏, ♓) with Venus—a powerful condition for two benevolent planets.

It is apparent that Jupiter played the central role in Augustus's horoscope. Moreover, Dorotheus of Sidon tells us the following about the Sun in the Ascendant and Jupiter in the Midheaven:

> If the Sun is in the Ascendant or Midheaven in its own house or a male sign, then it will be good. If the benefics [i.e., Jupiter] aspect it from the tenth place, [Midheaven], he will be praiseworthy, a leader over groups in nobility. (*Pentateuch* 2.22.1)

In any case, it is apparent that Augustus owed much of his imperial fate to Jupiter, which was exalted, in the Midheaven, in opposition with the night-ruling Moon, and in trine with co-ruling Venus.

Hadrian

Hephaestion of Thebes collected (ca. A.D. 415) a number of astrological charts, and one that he recorded from the works of Antigonus of Nicaea is without a doubt the chart of the emperor Hadrian, who was born January 24, A.D. 76, and adopted as heir to the imperial throne by his cousin, the emperor Trajan (see figure 12). In the horoscope recorded by Antigonus there are (minor) discrepancies between modern computations of the planetary positions and those that he reported.[13] Still, the key point is not the accuracy of Antigonus's astronomical computations but rather his astrological reasoning and his analysis of Hadrian's life, which provide invaluable insight into ancient astrological interpretations.

Here is Antigonus's analysis of why Hadrian became emperor:[14]

He became emperor, because of the presence of two planets [Jupiter ♃ and the Moon ☽] in the horoscopic degree [Horoscopus], and especially because the Moon was on the Ascendant which coincides with the horoscopic degree, and because Jupiter was about to rise in its morning phase [in the east] within seven days. And because of [the Moon's] attendant planets being in their own houses, and of them Venus, ♀, being in the exaltation [hypsoma] of her own house [Pisces ♓], and Mars, ♂, in his own triangle [Cancer ♋, Scorpius ♏, Pisces ♓] . . . , while both planets [Venus and Mars] are close together and about to rise soon after the Moon. Moreover, the Sun, ☉, the cosmos ruler, is also the Moon's attendant . . . , and the Sun himself is in turn attended by Saturn ♄, in his own house [Capricorn ♑] and by Mercury, ☿, both of them being in their morning rising. It remains to be shown that the Moon too was about to be in conjunction with a bright fixed star in the twentieth degree. For one must not only pay attention to the conjunction of the Moon with the planets, but also with fixed stars. (Antigonus of Nicaea)[15]

Antigonus's analysis of Emperor Hadrian's horoscope focuses on Jupiter. First, Jupiter and the Moon were in a very close conjunction (same degree) and in the Ascendant, specifically on the eastern horizon. His emphasis of this aspect points to the importance of Jupiter's being in close conjunction with the Moon. One of the conditions that Antigonus points out relates to the Magi's account of their star: Jupiter was about to make its morning appearance in the east (become a morning star) within seven days. Antigonus also explains the importance of the attendant planets. Besides all of this, Antigonus emphasizes that it was Jupiter that gave Hadrian imperial blessings. Here is a summary of the important astrological conditions of Hadrian's horoscope:[16]

- Exaltations: Venus was exalted in Pisces.
- Trines: Mars rules its trine, Cancer ♋, Scorpio ♏, Pisces ♓.
- Cardinal points: The three important celestial bodies, the Sun, the Moon, and Jupiter, were in the Ascendant, the most important cardinal point.
- Attendants: Jupiter, Mercury, and Saturn preceded the Sun. Venus and Mars followed the Moon. In addition, Antigonus says the Sun attended (by closely following) the Moon.
- Jupiter: The regal planet was in the Ascendant and in close conjunction with the Moon. It was about to become a morning rising planet, preceding the Sun as a protective attendant.

What Hadrian lacked in influences from the trines he made up with attendant planets, the "spear-bearers." And the influence of Jupiter, which was about to become a morning star, was of paramount importance. As Antigonus stressed, Hadrian received the worship of the people by the grace of Jupiter, particularly Jupiter's proximity with the Moon. Antigonus also mentioned the influence of a bright star.[17]

Chronology

ca. 1000 B.C.	*Babylonian astrologers began to watch the skies for omens to advise the king.*
ca. 750 B.C.	*Constellations that will become the zodiacal signs first appear in Babylonian documents.*
ca. 650 B.C.	*Babylonian almanacs list planetary movements for the year.*
626 B.C.	*Chaldeans attain control of Babylonia.*
587 B.C.	*Babylonians (Chaldeans) destroy Jerusalem and bring captured Jews to Babylonia.*
539 B.C.	*Persians invade Babylonia and free the captive Jews.*
521 B.C.	*Zoroastrian priests, called magi, fail to seize the Babylonian throne from King Darius.*
ca. 410 B.C.	*Natal horoscopes make their appearance in Babylonia. The twelve signs of the zodiac are standardized.*
330 B.C.	*Alexander the Great conquers Babylonia.*
323 B.C.	*Alexander the Great dies and leaves the Seleucid Empire controlling Babylonia.*
248 B.C.	*Parthians conquer the Seleucid Empire.*
237 B.C.	*The Decree of Canopus by King Ptolemy III of Egypt institutes the first calendar with a leap year.*
205 B.C.	*Extensive calculated tables of planetary and lunar positions are found in Babylonian documents.*
162–126 B.C.	*Hipparchus in Rhodes greatly advances Hellenistic astronomy.*
146 B.C.	*Rome conquers Greece, which opens the East to further Roman conquests.*

63 B.C.   *Octavian, who will become Augustus Caesar, is born; and Pompey expands Roman territory in the East and conquers Jerusalem.*

48 B.C.   *Judean troops under Antipater (Herod's father) help to relieve Julius Caesar besieged in Alexandria. The appointment of Herod to the governorship of Galilee by a grateful Caesar marks the start of a great career.*

46 B.C.   *Julius Caesar resynchronizes the calendar with the seasons and applies the leap-year system—the start of the Julian Calendar.*

44 B.C.   *Julius Caesar's assassination sends the Roman world into turmoil.*

42 B.C.   *Avenging Caesar's death, Mark Antony and Octavian (the future Augustus Caesar) defeat Brutus and Cassius at Philippi. Antony throws his support behind Herod to advance Roman (and his) interests in the East.*

41 B.C.   *Cleopatra begins her liaison with Mark Antony.*

40 B.C.   *Antony and Octavian support King Herod of Judea's attempts to fend off Parthian intrusions into Syria and Judea.*

37 B.C.   *With Roman support Herod regains control of Judea.*

31 B.C.   *Octavian defeats Antony and Cleopatra at Actium. Herod is summoned to Rhodes to account for his support of Antony, but Octavian forgives Herod and expands his kingdom.*

27 B.C.   *Octavian becomes Augustus Caesar, the first Roman emperor. He issues coins bearing Capricorn, his birth sign.*

22 B.C.   *Herod announces the start of the renovation of the Temple of Solomon and Temple Mount.*

6 B.C.   *Jesus is born on April 17. Tiberius seeks refuge on Rhodes, where he is trained in astrology by Thrasyllus. Quirinius goes frequently to Rhodes to visit Tiberius, who is shunned by most Romans.*

4 B.C.	*Herod dies. Augustus Caesar makes Archelaus ethnarch of Judea.*
A.D. 2	*Augustus summons Tiberius back to Rome from Rhodes. Tiberius brings the astrologer Thrasyllus with him.*
ca. A.D. 5–14	*Antioch issues coinage bearing Aries, the Ram.*
A.D. 6	*Archelaus is deposed and sent into exile. Quirinius becomes the legate of Syria and is put in charge of the annexation of Judea.*
A.D. 11	*Augustus Caesar publishes his horoscope to stem rumors of his imminent death. Astrologers are forbidden to predict anyone's death.*
A.D. 14	*Augustus Caesar dies and Tiberius becomes emperor.*
ca. A.D. 15	*Manilius finishes writing his didactic poem on astrology called* Astronomica
A.D. 26–36	*Pontius Pilate is prefect of Judea.*
A.D. 30–33	*Jesus is crucified.*
A.D. 36	*Tiberius's astrologer, Thrasyllus, dies.*
A.D. 37	*Tiberius dies and Caligula becomes emperor.*
A.D. 41	*Caligula is assassinated, and Claudius is made emperor by the Praetorian Guard. Claudius makes Herod Agrippa king of Judea.*
A.D. 44	*Herod Agrippa dies, and Claudius reinstitutes firm Roman control of Judea through the procurators.*
A.D. 50–90?	*The astrologer Dorotheus of Sidon compiles his* Pentateuch.
A.D. 52	*The governor of Syria, Quadratus, sends the procurator Cumanus back to Rome for banishment by Claudius, which postpones the inevitable revolt.*
A.D. 54	*Claudius dies and Nero becomes emperor.*
A.D. 55–58	*Quadratus in Antioch reissues and modifies the Aries coinage.*
A.D. 59	*Balbillus returns to Rome from Alexandria and becomes Nero's astrologer.*

A.D. 63   *Renovations of the Temple of Solomon and Temple Mount are completed.*

A.D. 64   *Balbillus instigates the slaughter of surrogates to save Nero from a comet. Nero blames the Christians for the fire that devastated Rome; his charges lead to the Great Persecution of Christians.*

A.D. 66   *Judea revolts against Rome.*

A.D. 68   *Nero commits suicide.*

A.D. 68–69   *Civil war produces several short-term emperors: Galba, Otho, and Vitellius. Vespasian leaves the war in Judea to his son, Titus, and wins the imperial throne.*

A.D. 70   *Jerusalem falls to Titus.*

A.D. 73   *Masada, the last major Jewish stronghold, falls to the Romans.*

A.D. 79   *Vespasian dies and Titus becomes emperor.*

A.D. 80–90   *The Gospels of Matthew and Luke are written, most likely in Antioch.*

A.D. 132–35   *Judea erupts in a second revolt, under Simon Bar Kochba. Jews are expelled from Jerusalem.*

ca. A.D. 150   *Claudius Ptolemy writes a number of important works including the* Tetrabiblos, *a primary source on ancient Greek astrology.*

ca. A.D. 180   *Antigonus of Nicaea records important horoscopes such as that of Hadrian. Vettius Valens of Antioch writes the* Anthology, *an important work on Greek astrology.*

A.D. 218–22   *Elagabalus from Emesa near Antioch is emperor and institutes the Sun-worshipping cult of Sol Invictus Elagabal. Aries, the Ram, appears again on Antioch's coinage.*

A.D. 224   *Parthia is conquered by Persians (Sasanians), who then threaten the eastern Roman provinces.*

A.D. 247–49   *Aries, the Ram, is used extensively on Near Eastern coinage.*

A.D. 256   *Antioch temporarily falls to the Sasanians.*

A.D. 260   *The emperor Valerian is captured by the Sasanians in Mesopotamia. Eastern city coinage falls off and eventually ceases.*

A.D. 275   *The emperor Aurelian decrees December 25 as the birthday of the Unconquerable Sun. Christians give this celebration a Christian meaning by celebrating Jesus' birth on the same day.*

A.D. 294   *Emperor Diocletian ceases all city (Greek imperial) coinage and standardizes controls on coinage. 4th and 5th centuries Several important astro-logical works are produced: Firmicus* Mathesis *(ca.* A.D. *334); Paulus Alexandrinus* Introduc-tory Matters *(ca.* A.D. *378); Anonymous of 379* The Treatise on the Bright Fixed Stars; *Hephaes-tion of Thebes* Apotelesmatics *(ca.* A.D. *415).*

A.D. 533   *Dionysius Exiguus proposes the Christian Era sys-tem for counting the years.*

Notes

CHAPTER 1. INTRODUCTION

1. Although the Protoevangelium of James mentions the Star of Bethlehem, it was written in the second century and is likely a combination of the Gospels of Luke and Matthew. For this reason, many biblical scholars claim that it is not an independent source confirming the Magi's star. See Raymond E. Brown, *The Birth of the Messiah. A Commentary on the Infancy Narratives in the Gospels of Matthew and Luke* (New York: Doubleday, 1993), 33, 176.

2. Michael R. Molnar, "The Coins of Antioch," *Sky and Telescope* 83, no. 1 (1992): 37–39; and Molnar, "The Magi's Star from the Perspective of Ancient Astrological Practices," *Quarterly Journal Royal Astronomical Society* 36 (1995): 109–26.

3. George MacDonald, "The Pseudo-Autonomous Coinage of Antioch," *Numismatic Chronicle,* 4th ser., 4 (1904): 105–35.

4. These denarii, RIC-1 541–2, are believed to have been issued after 27 B.C. The notation RIC-1 541–2 refers to coin nos. 541–2 in C.H.V. Sutherland, *The Roman Imperial Coinage* (London: Spink, 1984), vol. 1. The volumes are chronologically arranged according to the succession of emperors.

5. Kevin Butcher, *Roman Provincial Coins: An Introduction to the Greek Imperials* (London: Seaby, 1988), 97.

6. Mark, written in the late sixties, was the first gospel ("Markan priority"), and John was written in the nineties. See Brown, *Birth of the Messiah,* 27 n. 5. For an intriguing analysis of the dating of the crucifixion, see Bradley E. Schaefer, "Dating the Crucifixion," *Sky and Telescope* 77, no. 4 (1989): 374. The evangelists are discussed in Brown, *Birth of the Messiah,* 235. "Q" stands for the German word *Quelle* (source); see Brown, *Birth of the Messiah,* 45, 236. There are two other New Testament sources: "M," the primary source for Matthew, and "L," the source for Luke.

7. Kim Paffenroth, "The Star of Bethlehem Casts Light on Its Modern Interpreters," *Quarterly Journal Royal Astronomical Society* 34 (1993): 449–60.

8. Brown, *Birth of the Messiah,* 190–96.

9. Ibid., 36–37, 557–63.

10. The planning for the new temple started in 22 B.C., but the work did not start until two years later, when all correct religious customs were agreed upon and put in place. The inner temple itself was completed in one and one-half years, but the outer courts and ancillary buildings that make up the Temple Mount were still undergoing renovations during the time of the procurator Albinus (A.D. 62–64).

11. See Peter Richardson, *Herod: King of the Jews and Friend of the Romans*

(Columbia: University of South Carolina Press, 1996), 174–202, for Herod's ambitious building program.

12. Macrobius *Saturnalia* 2.4.11 is cited in Brown, *Birth of the Messiah*, 236. In Greek this anecdote is a pun. The word for "pig" is *hus,* and the word for "son" is *huius.*

13. Neil Asher Silberman, "Searching for Jesus: The Politics of First-Century Judea," *Archaeology* 47, no. 6 (1994): 31–40.

14. I show the NRSV along with the KJV to illustrate how interpretations and translations of the English text can differ. Only because most people are more familiar with the KJV do I cite that version throughout this work. However, it is useful to explore the differences between the major versions given in John R. Kohlenberger III, ed., *The Precise Parallel New Testament* (New York: Oxford University Press, 1995).

15. Suetonius *Vespasian* 4:5; Tacitus *Histories* 5:13; Josephus *Wars of the Jews* 6:6.4.

CHAPTER 2. A MYSTERIOUS STAR

1. For miracle theories, see D. C. Allison, "What Was the Star That Guided the Magi?" *Bible Review* 9, no. 6 (1993): 20–24, 63; B. T. Viviano, "The Movement of the Star, Matt 2:9 and Num 9:17," *Revue Biblique* 103 (1996): 58–64. For planets as messengers of God, see Lester J. Ness, "Astrology and Judaism in Late Antiquity," *The Ancient World* 26, no. 2 (1995): 126–33.

2. Colin J. Humphreys, "The Star of Bethlehem—A Comet in 5 B.C.—And the Date of the Birth of Christ," *Quarterly Journal Royal Astronomical Society* 32 (1991): 389–407.

3. The Ides of March prophecy of Julius Caesar's death was self-fulfilling because his enemies took advantage of the well-publicized omen. For this reason personal horoscopes were kept private. See Michael R. Molnar, "Astrological Omens Commemorated on Roman Coins: The Ides of March," *The Celator* 8, no. 11 (1994): 6–10; and Molnar, "Astrological Omens Commemorated on Roman Coins: Clues to Caesar's Fortune," *The Celator* 10, no. 3 (1996): 14–19. For the Julian comet, see John T. Ramsey and A. Lewis Licht, *The Comet of 44 B.C. and Caesar's Funeral Games* (Atlanta: Scholars Press, 1997).

4. Tacitus *Annals* 15, 47–65; and Dio Cassius *Roman History* 62, 24, 1.

5. Frederick H. Cramer, *Astrology in Roman Law and Politics* (Philadelphia: American Philosophical Society, 1954; reprint, Chicago: Ares Publishers, 1996), 118–28.

6. Mithridates the Great (ca. 134–63 B.C.) of Pontus (northern Turkey) is another person who more or less successfully used cometary appearances to his personal gain. For this reason, proponents for a "benign" comet as the Magi's star cite Justin *Epitome of the Philippic History of Pompeius Trogus* 37.2.1; however, in "Mithradates Used Comets on Coins as a Propaganda Device," *The Celator* 11, no. 6 (1997): 6–8, I show that Mithridates' comet was an evil hippeus comet that foretold the overthrow of the tyrannical Romans. The foreboding interpretation of cometary portents is given in A. A. Barrett, "Observations of Comets in Greek and Roman Sources before A.D. 410," *Journal Royal Astronomical Society Canada* 72, no. 2 (1978): 81–106. Hephaistio [Hephaestion] of Thebes *Apotelesmatics* 1.24 lists ten types of comets; all are evil portents except for the Comet of Zeus and only if

it is in Cancer, Scorpio, or Pisces. This minuscule exception is hardly an argument in favor of benign comets.

7. Until modern times there was no distinction between a nova and a supernova.

8. Max Caspar, *Kepler* (London: Abelard-Schuman, 1959), 154–59. For the idea that Kepler's astrology was modern in character, see A. Pannekoet, *A History of Astronomy* (New York: Dover, 1989), 234–44.

9. Caspar, *Kepler,* 153.

10. All translations of *Kepleri opera omnia* are from W. Burke-Gaffnet, "Kepler and the Star of Bethlehem," *Journal Royal Astronomical Society Canada* (December 1937): 417–25.

11. Gary Kronk, "A Large Comet in 135 B.C.," *International Comet Quarterly* (January 1997): 3–7. Kronk argues persuasively that the "new star" of Hipparchus was not a comet as some other researchers believe.

12. David H. Clark, John H. Parkinson, and F. Richard Stephenson, "An Astronomical Re-appraisal of the Star of Bethlehem—A Nova in 5 B.C.," *Quarterly Journal Royal Astronomical Society* 18 (1977): 443–49. Note that this paper is an astronomical interpretation, not an astrological one. The Chinese record is discussed in Ho Peng-Yoke, "Ancient and Mediaeval Observations of Comets and Novae in Chinese Sources," *Vistas in Astronomy* 5 (1962): 218.

13. Roger W. Sinnott, "Thoughts on the Star of Bethlehem," *Sky and Telescope* 36, no. 6 (1968): 384–86.

14. Burke-Gaffnet, "Kepler." This article is a fountain of information on Kepler's work involving the Magi's star.

15. I translated this passage from the French text: Vettius Valens *Vettius Valens d'Antioche—Anthologies* (trans. Bara) bk. 1. However, all other references to the *Anthology* are from the English translation listed in the bibliography.

16. Valens *Anthology* 1.20.

17. Here we apply astronomy historian Otto Neugebauer's finding that at the beginning of the Christian Era astrological calculations used approximately the fifth degree of Aries as the vernal equinox (Υ 5° = v.e.; see Appendix A). The modern convention of defining the position of the zodiacal signs, namely, that Aries started at the vernal equinox (Υ 0° = v.e.), is incorrect for ancient horoscopes. (For those who insist on using the modern convention, Mars joined Jupiter and Saturn in Pisces on January 22, 6 B.C. Then Jupiter left Pisces and entered Aries on approximately February 28.)

18. The rabbi is also sometimes called Abrabanel.

19. David Hughes, *The Star of Bethlehem: An Astronomer's Confirmation* (New York: Walker and Company, 1979), 68, 184–86.

20. Tamsyn Barton, *Ancient Astrology* (New York: Routledge, 1994), 54. Of course, the evidence from O. Neugebauer and H. B. Van Hoesen, *Greek Horoscopes* (Philadelphia: The American Philosophical Society, 1987), is that natal charts were preferred over conception charts. There was a minor school of thought that held that the time of conception was important, but defining that hour necessitated asking delicate questions and invading people's privacy and proved impossible.

21. Abarbanel's astrological views were not shared by conservative Jews of Herod's time, although Hellenistic Jews gave astrology a Jewish interpretation.

22. Although astrologers' assignment of countries to zodiacal signs varies especially in late antiquity, none equated Pisces to Judea. The only reference specifically to Judea is in Ptolemy *Tetrabiblos* 2.3, in which Ptolemy mentions Judea twice. His assignment is quoted by other astrologers, such as Vettius Valens. See Matt. 17:27, John 21:1–14, and Mark 6:34–44 for the biblical stories about fishes and Jesus.

23. A trine is also called a triplicity, trigon, or triangle. Ptolemy assigned the four trines to the winds from the cardinal directions: Trine I was the northwest wind, Trine II the southeast, Trine III the northeast, and Trine IV the southwest. See *Tetrabiblos* 1.18. Rhetorius the Egyptian of the sixth century A.D. is credited with assigning the trines to the four elements, fire, earth, air, and water, respectively. See Barton, *Ancient Astrology*, 82.

24. Hughes, *Star of Bethlehem*, 122–30.

25. Tacitus *Histories* 5.4. The only corroborating evidence in Greek astrology for Saturn's purported connection to Jews is that Saturn was one of three celestial bodies ruling over the sign of Judea, namely, Aries. Saturn was the common sect ruler of the trine of Aries, Leo, and Sagittarius. Furthermore, Aries was the domicile, or house, of Mars, which Ptolemy noted as purportedly bestowing certain characteristics on the Jews. See *Tetrabiblos* 2.3. Note that Aries was the fall, or depression, of Saturn, which meant that Saturn was weak in that sign.

26. Being in the east at sunset is an acronychal rising, which is not an auspicious condition. More important, when a planet was "in the east" it was not in an acronychal rising but becoming a morning star in the eastern sky. According to Stoic beliefs that played an important role in Greek astrology, a planet emerging from the fiery region of the Sun was undergoing *ekpyrosis* and *palingenesis*, a cyclical fiery death and rebirth. A planet reborn as a morning star was in its strongest phase of astrological influence.

27. Hughes, *Star of Bethlehem*, 68, 184–86.

28. P.A.L. Chapman-Rietschi, "Venus as the Star of Bethlehem," *Quarterly Journal Royal Astronomical Society* 37 (1996): 843–44. The theory about Venus also draws from Sumerian and Babylonian mythologies.

29. Hughes, *Star of Bethlehem*, 178–81.

30. O. Neugebauer, *The Exact Sciences in Antiquity* (New York: Barnes & Noble, 1993), 167–71.

CHAPTER 3. THE DAWNING OF THE FIRST MILLENNIUM

1. For the origins of *magi*, see Herodotus *The History* 3.61–80, 7.37. For indications that magi were esteemed advisors, see Brown, *Birth of the Messiah*, 193–96. For the belief that magi were considered as royalty, see Tertullian *Adversus Marcion* 3.13.

2. See the Amplified Bible, New American Standard Bible, and New Revised Standard Version. The references to the Magi can be compared in Kohlenberger, *Precise Parallel New Testament*, 6–9.

3. Reason (λόγος = *logos*) is a fundamental, guiding principle of Stoic philosophy. *Logos* also meant "a thought," "a word," or sometimes "divinity." Logical knowledge is a virtue, and reason is a divine force that manifests itself as Fate. Destiny is thus believed to be divine pure reason. M. Carey et al., eds., *The Oxford Classical Dictionary* (London: Oxford University Press, 1950), c.v. "stoa."

4. Luther M. Martin, *Hellenistic Religions: An Introduction* (Oxford: Oxford University Press, 1987), 43–44. For a thorough review of astrology's synergy with Roman religion, see Franz Cumont, *Astrology and Religion among the Greeks and Romans* (New York: Dover, 1960). Few books on Roman history mention the influence of astrology. For an excellent treatment of astrology in Roman culture, see Michael Grant, *The World of Rome* (New York: Penguin Books, 1960), 149–76.

5. Ness, "Astrology and Judaism." There is evidence that some Jews eventually adapted astrology to their monotheism by placing the planetary influences subordinate to God. For a good review of Jewish attitudes toward astrology, see Barton, *Ancient Astrology*, 68–70. That Jews did not practice astrology is discussed in W. E. Mills, ed., *Mercer Dictionary of the Bible* (Macon: Mercer University Press, 1990), 73. See also Deut. 4:19 and Isa. 47:13.

6. The Wise Men brought three gifts: gold, frankincense and myrrh. Although gold is still recognized as a valuable metal, the other gifts are little known today. Some people may know frankincense for its aromatic fragrance, but in ancient times it was prized for its medicinal uses in treating wounds and parasitic infections. Myrrh, another medicine, was valued primarily as a powerful antiseptic and painkiller but was also used as an antifungal agent for embalming corpses. The value of their offerings suggests to some people that the Magi may even have been kings, but actually the gifts are typical for prosperous physicians—the art of healing was another specialty of magi.

Nowhere is the number of Wise Men specified, but the lore is that because they brought three gifts there must have been three men. As the story about the Wise Men was embellished over the centuries, they received names: Balthazar, Melchior, and Gaspar. However, there are other stories claiming that they numbered from two to as many as fourteen. See Hughes, *Star of Bethlehem*, 25.

7. Brown, *Birth of the Messiah*, 168–70. Arabia is another candidate for the origin of the Magi; however, Hughes (*Star of Bethlehem*, 35) makes a compelling argument for the lands under Parthian control (Babylonia and Mesopotamia). That the Magi may have come from Parthia is an intriguing possibility. In 587 B.C. the Babylonians overran Judea and transported many captured Jews back to Babylonia. Cyrus, the Persian, conquered Babylonia in 539 B.C. and freed the Jews, but many chose to remain rather than return to Judea. The Parthians took over Babylonia in ca. 248 B.C., and under Parthian control Babylonian Jews prospered. Moreover, some Parthians, even among the nobility, converted to Judaism, and these conversions strengthened the bond between Judea and Parthia.

There were times when Parthians tried to assist Jews in fending off the Romans, and vice versa, times when Jews supported Parthian interests. The Parthians supported Mattathias Antigonus (the last Hasmonean king of Judea) in 40 B.C., but the Romans ousted him in 37 B.C. and put Herod on the throne. Well after the time of Herod, Jews continued to look to Parthia for release from Roman tyranny, which is related in a story about Rabbi Simeon Ben Yohai, a leader in the second revolt (A.D. 132–35), who said that the Parthians would accompany the Messiah in liberating Judea: "If you see a Persian [Parthian] horse tethered to a grave in Israel, look for the footsteps of Messiah." See Jacob Neusner, *A History of the Jews in Babylonia*, vol. 1, *The Parthian Period* (Leiden: E. J. Brill, 1965), 79. (The Parthian horsemen were legendary fighters. See Fred B. Shore, *Parthian Coins and History: Ten Dragons against Rome* [Quarryville, Penn.: Classical Numismatic Group, 1993], 69.)

If the Magi were from Parthia as many people believe, their appearance in Herod's court would have been an undeniable threat to Herod. The possibility that Parthia, Herod's nemesis, could throw its support behind the Messiah goes a long way toward explaining why Herod "was frightened." Even if the Magi had not been from Parthia, Herod would have realized that Parthian astrologers practicing Greek astrology also recognized what the Magi had seen; that is, he would have expected that his enemies also knew about the Messiah's star.

8. For information on the gravestone (tablet) in figure 4, see also Ruth S. Freitag, *The Star of Bethlehem: A List of References* (Washington, D.C.: Library of Congress, 1979), 44. For a discussion on the Phrygian cap worn by the Magi in figure 4, see David Ulansey, *The Origins of the Mithraic Mysteries: Cosmology and Salvation in the Ancient World* (New York: Oxford University Press, 1989), 26–28. See also Grant, *World of Rome,* 201, for remarks about the star-studded Phyrgian cap worn by the cosmic deity Mithras.

9. George Huxley, *The Interaction of Greek and Babylonian Astronomy* (Belfast: The Queen's University, 1967), 1; Cramer, *Astrology in Roman Law,* 9–19; O. Neugebauer, *Astronomy and History: Selected Essays* (New York: Springer-Verlag, 1983), 60–87.

10. Anonymous, quoted in Judith Kingston Bjorkman, "Meteorites in the Ancient Near East," *Meteoritics* 8, no. 2 (1973): 91–132.

11. There is evidence that a few constellations—such as Ursa Major, the Great Bear—may have originated as far back as the Ice Ages, ca. 10,000 B.C. The vast majority of constellations in Western culture evolved over time. Owen Gingerich, *The Great Copernicus Chase* (Cambridge, Mass.: Sky Publishing Corp. and Cambridge University Press, 1992), 1–12.

12. The first horoscopes referred to the risings, settings, and other transitions of celestial bodies that figured as important events in the infant's future. See Barton, *Ancient Astrology,* 16. For the origins of the zodiacal signs, see Owen Gingerich, "The Origin of the Zodiac," *Sky and Telescope* 67, no. 3 (1984): 218–20; Jim Tester, *A History of Western Astrology* (New York: Ballantine, 1987), 14.

13. F. E. Peters, *The Harvest of Hellenism: A History of the Near East from Alexander the Great to the Triumph of Christianity* (New York: Barnes & Noble, 1996), 111–12.

14. Neugebauer, *Exact Sciences,* 105. The Babylonian mathematical techniques are recorded in the section about "procedure texts." See also Hugh Thurston, *Early Astronomy* (New York: Springer-Verlag, 1994), 64–81; Barton, *Ancient Astrology,* 19; Neugebauer, *Astronomy and History,* 33–142, 157–64.

15. Neugebauer, *Astronomy and History,* 112–16. A good review of Greek contributions to mathematical astronomy is given by Thurston, *Early Astronomy,* 110–77.

16. There were some "Egyptian" contributions, too, but they were usually from Hellenistic Egyptians; see Barton, *Ancient Astrology,* 23. The argument for the Hellenistic nature of ancient astrology is forcefully developed by Francesca Rochberg-Halton, "New Evidence for the History of Astrology," *Journal of Near Eastern Studies* 43, no. 2 (1984): 115–40; see also her "Elements of the Babylonian Contribution to Hellenistic Astrology," *Journal of the American Oriental Society* 108, no. 1 (1988): 51–62.

17. That astrologers were calculating the positions of celestial bodies is evi-

dent from the collected works of O. Neugebauer. See also Hughes, *Star of Bethlehem*, 128.

18. Astrologers could "predict" eclipses and occultations only to a certain extent. Parallax effects, in particular, prevented definite predictions. Thus, astrologers knew when an eclipse could not happen and when one was likely to happen, but they could not definitely predict that one *would* happen.

19. Caesar was assassinated in 44 B.C. just before he was scheduled to leave for the campaign against Parthia. Years later, in 31 B.C., Agrippa would lead Augustus's naval forces to victory against Mark Antony and Cleopatra at Actium.

20. There are many examples of life expectancy studies; see Neugebauer and Van Hoesen, *Greek Horoscopes*, 87–89, literary horoscope L75.

21. This procedure is essentially harmonic analysis and perturbation theory. For an overview of the procedure, see Neugebauer, *Astronomy and History*, 47–50.

22. The accuracy of ancient horoscopes is difficult to gauge because of differences in definitions of celestial positions. Neugebauer and Van Hoesen, *Greek Horoscopes*, 170–74, 180–84, 188.

23. A zodiacal sign is a fixed sector thirty degrees in extent that does not coincide exactly with the constellation (see Appendix A).

24. This discussion applies to the northern hemisphere, where the Greeks and Babylonians lived.

25. According to modern astronomers the constellation of the Scorpion is called Scorpius, but astrologers call the sign Scorpio.

26. The fact that Ptolemy stresses the importance of the Ascendant and Midheaven while virtually ignoring the rest of the system of houses indicates an early, elementary form of astrology that was to evolve in the writings of later Greek documents. See Franz Boll, *Studien über Claudius Ptolemäus* (Leipzig: B. G. Teubner, 1894), for evidence of the early origin of the *Tetrabiblos*. Note that Ptolemy also presented his own improvements of old systems, for example, the system of terms.

27. Boll (*Studien,* 181–238) claims this list of countries comes from Posidonius (first century B.C.).

28. By "godless," Ptolemy means that Jews did not worship the pagan pantheon. Furthermore, the night domicile of Mars is Aries. Domiciles were of relatively minor importance in regal horoscopes; thus, we should not expect that Mars, the domicile ruler of Aries, had to be a component in the portent that constituted the Star of Bethlehem. There are, however, exceptions, times when domiciles are mentioned: Antigonus referred to them in Hadrian's horoscope; however, they played minor roles in the regal analysis. For Hadrian's horoscope, see Cramer, *Astrology in Roman Law,* 165–70, and Appendix D.

29. Ptolemy put Syria under the control of Scorpio:

The people of Syria, Commagene, and Cappadocia are familiar to Scorpio and Mars; therefore much boldness, knavery, treachery, and laboriousness are found among them. (*Tetrabiblos* 2.3)

See RPC-1 3852–4, 3856. These are coin nos. 3852–4 and 3856 in Andrew Burnett, Michel Amandry, and Pere Pau Ripollès, *Roman Provincial Coinage* (London: British Museum, 1992), vol. 1. For the use of Scorpio as symbolizing Commagene, see

Tetrabiblos 2.3. The evidence suggests that the scorpion symbolizes the zodiacal sign and is not a depiction of the commonly found animal. Commagene also issued coinage with another astrological sign—namely, Capricorn, on 3855 and 3861–2—most likely in deference to Augustus and Vespasian, who identified themselves with that sign. The royal family of Commagene was linked by marriage to the influential and esteemed astrologer Thrasyllus, who may have influenced the use of astrological signs on Commagene's coinage. See Cramer, *Astrology in Roman Law*, 94, for Thrasyllus's marriage.

30. Valens *Anthology* 1.2.

31. Manilius *Astronomica* 4.744–54.

32. In his *Introduction to Astrology* (1.9), Arab astrologer Abu Ma'šar (A.D. 787–886) assigned "Babylonia, Persia, Azerbaijan and Palestine" to the sign of Aries, which indicates that the connection between Aries and Palestine was still remembered well after Roman times.

33. Antioch was a Hellenistic city in the sense that it was Seleucid and looked toward Greece and Rome for its culture. It was nevertheless quite Semitic. Aramaic was the chief language. Stewart Perowne, *Herod* (New York: Dorset, 1991), 126. That Antioch played important roles in Roman and Christian histories is summarized by Matthew Bunson, *A Dictionary of the Roman Empire* (New York: Oxford University Press, 1995), 16.

34. Michael Grant, *Roman History from Coins: Some Uses of the Imperial Coinage to the Historian* (New York: Barnes & Noble, 1995), 78–81. This local coinage died in the mid–third century A.D., when the empire was momentarily fractured by civil wars

35. Josephus *Antiquities of the Jews* 17.13. Roman Vienna in Gaul is immediately south of modern Lyon, France, on the Rhone River.

36. At that time Quirinius was conducting a census for taxation in Syria. See Fergus Millar, *The Roman Near East: 31* B.C.–A.D. *337* (Cambridge: Harvard University Press, 1993), 48. Tacitus (*Annals* 2.42.7) reports that in A.D. 17 Syria and Judea petitioned Rome for a reduction in the tribute. A good review of the political and religious affairs of Judea is given by Robert M. Grant, *Augustus to Constantine: The Emergence of Christianity in the Roman World* (New York: Barnes & Noble, 1996), 24–39.

37. From 6 B.C. to A.D. 2, Tiberius lived isolated in exile on the island of Rhodes. If he had any close friends during that time, Quirinius (who frequently visited him) and Thrasyllus (who taught him astrology) must have been among them. It is possible that Thrasyllus and Tiberius involved Quirinius in their discussions of astrology, and thus Quirinius may have been familiar with astrology.

38. MacDonald ("Coinage of Antioch") remarked about this striking similarity between the coins' depiction of Aries under a star and the Lucan account of the shepherds tending their flocks at night. He mused about a possible relationship between the coin and the events of A.D. 6 mentioned in Luke 2:1–2 concerning the census of Quirinius and the birth of Jesus.

39. For the mythology about Aries, see Julius D. W. Staal, *The New Patterns in the Sky* (Blacksburg, Va.: McDonald and Woodward, 1988), 36–41.

40. City coinage always carried its *ethnic*, a word that names the citizens of the city in the possessive genitive plural; in this case the ethnic is translated as "Of the people of the city of Antioch." The designation of the Antiochene municipal

coinage is according to Burnett, Amandry, and Ripollès, *Roman Provincial Coinage*, 1:606–30. The civic series includes RPC-1 4265 and 4266. Because of the style of the undated Aries coin, MacDonald ("Coinage of Antioch") suggested it is from A.D. 5–10.

41. RPC-1 4268 and 4269.

42. Brown, *Birth of the Messiah*, 547–63.

43. E. A. Wallis Budge, *The Rosetta Stone* (New York: Dover Publications, 1989), 251–95.

44. W. M. O'Neil, *Time and the Calendars* (Parramatta, Australia: Sydney University Press and Macarthur Press, 1975), 77–87.

45. Note that there is no year A.D. 0 or 0 B.C. A.D. (anno Domini) stands for "in the year of the Lord," which is how the years in the Christian Era system are counted from the estimated time of Jesus' birth. Sometimes the designation C.E., Christian Era or Common Era, is used instead.

This work finds April 17, 6 B.C., as the most likely date for Jesus' birth. If we base the numeration of the millennia on the birth of Jesus, then the third millennium of the Christian Era has already begun. That is, if the first year, A.D. 1, were moved to the day the Star of Bethlehem appeared and a revised Christian Era calendar were recomputed by means of the convention established by Dennis Exiguus, then the year A.D. 2000 would have begun on April 17, 1994. By April 17, 1995, two thousand years had passed, which means that the millennium anniversary of his birth has already passed.

46. Clement of Alexandria *Stromata* 1.21.145.

47. Clement A. Miles, *Christmas Customs and Traditions: Their History and Significance* (New York: Dover Publications, 1976), 20.

48. The Roman religion of Sol Invictus is discussed in Gaston H. Halsberghe, *The Cult of Sol Invictus* (Leiden: E. J. Brill, 1972). According to Ulansey (*Mithraic Mysteries*, 109) Mithras was the *kosmokrator* of the sphere of fixed stars, and the Sun was the *kosmokrator* of the Moon and planets. This title meant a world ruler and denoted divine omnipotence.

49. Ernest L. Martin, *The Birth of Christ Recalculated* (Pasadena, Calif.: Foundation for Biblical Research, 1980); John Mosley, *The Christmas Star* (Los Angeles: Griffith Observatory, 1987).

50. The "crowded" sequence of events included the execution of rabbis for tearing down an eagle statue; Herod's travel from Jericho to the Dead Sea to find relief from his ailment; Herod's plan to assemble and kill Jewish elders; the execution of Herod's son Antipater; and finally the death of Herod. However, the majority of historians have argued that there was indeed sufficient time in 4 B.C. for everything to have taken place and that there is no historical evidence for moving Herod's death from 4 B.C.

51. Josephus *Antiquities of the Jews* 17.6.4.

52. Hephaistio of Thebes *Apotelesmatics* 1.21.

53. Ronald Syme, *The Roman Revolution* (London: Oxford University Press, 1963), 399 n. 4.

54. Brown, *Birth of the Messiah*, 547–63; Josephus *Antiquities of the Jews* 18.1.1.

55. Brown, *Birth of the Messiah*, 412–18.

56. The fifteenth regnal year of Tiberius in the Roman dating system spans

A.D. 27–28. Biblical scholars use the Jewish tradition of counting, which gives A.D. 28–29. The Roman system gives better agreement because it makes Jesus thirty to thirty-five at the start of his ministry.

57. Hughes, *Star of Bethlehem*, 70–75.

58. Brown, *Birth of the Messiah*, 395.

59. See Kohlenberger, *Parallel New Testament*, for The King James Version, Amplified Bible, Rheims New Testament, New American Standard Bible, New American Bible, New International Version, and New Revised Standard Version.

60. For the taxation system under Augustus, see A.H.M. Jones, *Augustus* (New York: Norton, 1970), 118–19. The rioting produced by the tax is in Josephus *Antiquities of the Jews* 18.1.1. Quirinius's taxation is discussed in Millar, *Roman Near East*, 35, 48.

61. Brown, *Birth of the Messiah*, 553. Brown also notes that Tertullian gives conflicting estimates for Jesus' birth. The reference to multiple censuses is, as Brown believes, a plausible explanation for a census before the one of Quirinius.

62. Millar (*Roman Near East*, 46–47) notes that it is not understandable why the family of Jesus would have had to leave Galilee for a census in Judea. Galilean Jews were protected from any Roman tribute during the entire period of Jesus' lifetime. Galilee was under the tetrarchy of Antipas (4 B.C.–A.D. 39) and was not controlled by the Roman prefect ruling Judea. Similarly, Judea and Samaria were exempt from any Roman tax enrollment during the reigns of Herod and Archelaus.

63. It is virtually impossible in the principles of Greek astrology for a regal portent to last for two years. However, it is likely that Herod was threatened by a series of separate regal portents.

64. Hughes (*Star of Bethlehem*, 75) concluded that the biblical evidence points to 8–6 B.C. for Jesus' birth.

Chapter 4. Secrets of Regal Horoscopes

1. All possible regal conditions and aspects are not covered, just the most important items. For example, the Lot of Fortune had very powerful regal implications, but its calculation depends on the exact time of birth, which is impossible to ascertain for the birth of Jesus.

2. There was a third component of the natal chart: the system of lots. This component, which combined concepts of the other two, applied angular differences between celestial bodies, especially the Moon and the Sun, and the Ascendant (Horoscopus) to derive powerful locations along the zodiac. For a discussion of the lots, see Neugebauer and Van Hoesen, *Greek Horoscopes*, 8–9; Ptolemy *Tetrabiblos* 2.10; and Manilius *Astronomica* 3.43–159.

3. Dorotheus (*Pentateuch* 1.23–25) uses the sign as the exaltation; Firmicus (*Mathesis* 2.3.4–6) explains that although being in the sign was good, the exact degree was most propitious. The exact degrees were the Sun in 19° Aries; the Moon in 3° Taurus; Mercury in 15° Virgo; Venus in 27° Pisces; Mars in 28° Capricorn; Jupiter in 15° Cancer; and Saturn in 21° Libra. Nevertheless, the evidence from the record of horoscopes is that the sign of the exaltation was used. See Barton, *Ancient Astrology*, 96–97, and especially Neugebauer and Van Hoesen, *Greek Horoscopes*, 7.

4. The planetary configuration for exaltations is partly fictitious. Mercury ☿, is never more than 27° from the Sun ☉, and in figure 13 it is shown 146° from the

Sun—a physical impossibility. Thus, the configuration cannot represent any specific day.

5. Ptolemy (*Tetrabiblos* 2.3) ignored Saturn, which other astrologers recognized as a co-ruler of the trine of Aries, Leo, and Sagittarius. He also introduced some co-rulers that were not recognized by other astrologers. Ptolemy expressed his personal creativity in other astrological concepts as well (for example, he set out his own system of terms and lots). In this work, the conventions adopted by the majority of astrologers are applied.

6. Note that a conjunction is a special case of the rulers of a trine being in the same sign. This fact may explain why the *Tetrabiblos* does not treat the conjunction as a unique aspect.

7. The cardines, or "centers," are four points defining the rising point (Horoscopus), setting point (*Dusis, Occasus*), upper culmination (*Medium Caelum*), and lower culmination (*Imum Caelum*). However, the astrologers usually referred to the more practical *loci* (*topoi* and also called "houses"), which are sectors based upon the location of the centers. See Appendix C.

8. The other important case of spear-bearing is when a planet is in aspect, especially in trine, to the Sun or Moon. It is also significant when the spear-bearer is in a cardine. See Hephaistio of Thebes *Apotelesmatics* 1.17.

9. Cramer (*Astrology in Roman Law*, 248–83) discusses the legal restrictions on astrology during the Roman Empire.

10. The trine aspect is best if the attendants lie clockwise ("in dexter aspect") with respect to these important cardines.

11. Pisces is a domicile of Jupiter, and the appearance of Jupiter in Pisces is thus a good portent, but one without regal importance. Pisces is a "feminine" sign and thus does not contribute to Jupiter's masculinity, or to a king's power. See the discussion about attendant planets and Appendix B.

12. For a fine treatment of "bright stars," see the translation and excellent commentary in Anonymous of 379 *The Treatise on the Bright Fixed Stars* (trans. Schmidt, ed. Hand). See also Firmicus *Mathesis* 6.2, 8.31.

13. For a discussion of the Nemrud Daği horoscope, see Neugebauer and Van Hoesen, *Greek Horoscopes,* 14–16. See also Bouché-Leclerq, *L'Astrolgie Grecque* (1899; reprint, Bruxelles: Culture et Civilisation, 1963), 439. The star is unquestionably Regulus near the lion's heart. The Sun was in Cancer at the time, which means that the star in the arms of the crescent moon could not be the Sun. For an alternative interpretation, see Donald H Sanders, ed., *Nemrud Daği: The Hierothesion of Antiochus I of Commagene* (Winona Lake, Ind.: Eisenbrauns, 1996), 89–90.

14. See Bradley E. Schaefer, "Heavenly Signs," *New Scientist,* December 21–28, 1991, 48–51; and Schaefer, "Origin of the Star and Crescent," *Sky and Telescope* 83, no. 2 (1992): 129. A close conjunction with the Moon and a planet is also a plausible interpretation of the star and crescent. But to be the powerful, good portent that the star and crescent symbol seems to portray, the conjunction must involve Jupiter or Venus. Venus bestowed graces involving love and family, but Jupiter gave regal powers and fame.

15. The royal bright star Regulus is in the trine of Aries, which added regal significance to aspects in this trine.

16. Michael Baigent, *From the Omens of Babylon: Astrology and Ancient Mesopotamia* (London: Arkana, 1994), 140–51.

17. Ivor Bulmer-Thomas, "The Star of Bethlehem—A New Explanation—Stationary Point of a Planet," *Quarterly Journal Royal Astronomical Society* 33 (1992): 363–74. This work stands as one of the best examinations of the Star of Bethlehem and explains the movements of the star and confirms that Jupiter was the regal star of the Magi.

18. Antigonus of Nicaea, quoted in Cramer, *Astrology in Roman Law,* 169. *Proskynesis* is the term that the biblical account in Matthew uses when Herod says he wants to "worship" the newborn king.

19. Hughes, *Star of Bethlehem,* 178–81.

20. Valens *Anthology* 1.19–20.

21. Modern calculations show that the Moon was not in the same degree as Jupiter at Hadrian's birth. Nevertheless, Antigonus's perception is more important than how close his calculations were to reality.

22. O. Neugebauer, *A History of Ancient Mathematical Astronomy* (New York: Springer-Verlag, 1975), 1038–51.

23. Astrologers also referred the "orb of separation," a zone around a celestial body that extends beyond its physical limits and thus extends its influence. The size of the orb of separation varied with each celestial body. Moreover, the values also varied with astrologers.

CHAPTER 5. ASTROLOGICAL PORTENTS FOR JUDEA

1. Originally I limited the time period for the search to 10 to 1 B.C. See Molnar, "Magi's Star."

2. Another astrological parameter was the position of the lunar nodes, the points at which the Moon crosses the ecliptic. See Neugebauer and Van Hoesen, *Greek Horoscopes,* 171. The Moon wanders up to five degrees above and below the ecliptic, but Jupiter stays within a little more than a degree of the ecliptic. Astrologers monitored the location of the lunar nodes to determine the likelihood of eclipses and other astrological effects. In the spring of 6 B.C. the ascending lunar node was in Aries, and thus the presence of Jupiter in Aries signaled the possibility of a lunar occultation of that planet. Like eclipses, occultations could not be predicted precisely; the astrologers of the first century B.C. knew only that an occultation was possible. Eclipses and occultations had to be verified by visual examination mainly because there was no way to handle the effects of parallax, the line of sight perspective that shifts the Moon in the sky. The likelihood of a lunar occultation of Jupiter could also be estimated by means of the fact that these occultations clustered in periods of about six years. For a discussion of occultations, see Bouché-Leclerq, *L'Astrologie Grecque,* 105.

3. Ptolemy *Almagest* 13.7–10. The visibility zone was called the *arcus visionis.*

4. Ptolemy *Tetrabiblos* 3.10; Firmicus *Mathesis* 2.9.1

5. Bulmer-Thomas, "The Star of Bethlehem."

6. Venus and Mercury also undergo retrograde motion and stationing.

7. Bulmer-Thomas, "The Star of Bethlehem."

8. Ptolemy *Tetrabiblos* 2.6.

9. The occultation of Jupiter by the Moon happened about an hour after Jupiter was at the Midheaven.

10. Aries was the fall, or depression, of Saturn. It is not obvious how astrolo-

gers would have treated this in the horoscope of April 17, 6 B.C. However, because Saturn, like Jupiter, was in a morning rising and was a co-ruler in its trine, any weakening effects were probably mitigated.

11. That there were many other houses or lots in which the regal nature was enhanced or preserved can be inferred from Valens's discussion in the beginning of chapter 2 of his *Anthology*.

12. Statistical investigations of the lunar occultation of April 17, 6 B.C., have been made. See M. M. Dworetsky and S. J. Fossey, "Lunar Occultations of Jupiter and Saturn, and the Star of Bethlehem," *The Observatory* 118, no. 1142 (1998): 22–24. Their analysis finds the occultation to be "rare," a conclusion based upon modern astronomy rather than Greek astrology, which, nevertheless, can be seen as reinforcing the importance of that day.

13. Tamsyn Barton, "Augustus and Capricorn: Astrological Polyvalency and Imperial Rhetoric," *Journal of Roman Studies* 85 (1995): 33–51. Researchers of ancient astrology use the term *polyvalency* to define the process that produced an auspicious, powerful horoscope.

14. Firmicus *Mathesis* 6.9.4–5. Mars was in quartile aspect with Saturn in Nero's chart, and he was predicted to lose his throne. See Suetonius *Nero* 40.

15. A biography of Firmicus appears in the introductory remarks in Firmicus *Mathesis* (trans. Bram) 1–4. See also Tester, *Western Astrology*, 133–42, for a discussion of the *Mathesis*. For the possibility that Abram cited in the *Mathesis* is the Jewish patriarch Abraham, see Firmicus *Mathesis* (trans. Bram) 311 n. 60. Bram claims that this reference could have been an attempt by Firmicus to ascribe astrological teachings to ancient wise men. Although there is a theory that Abram was not the patriarch but an unknown astrologer, one could nevertheless argue that Firmicus's placement of Abram among august notables such as Orpheus makes it more likely that Abram was in fact Abraham rather than an obscure astrologer. See Tester, *Western Astrology*, 133 n. 62, for Firmicus's switch between paganism and Christianity. As noted by Cramer (*Astrology in Roman Law*, 4), this transitional period between paganism and Christianity saw the emperor Constantine identifying himself with Helios even though he held Christian beliefs. For a description of Constantine's vision at Milvian Bridge, which was either of a cross or the Chi-Rho symbol that stood for Christ, see Robert M. Grant, *Augustus to Constantine*, 235–36.

16. Although this paragraph appears in a discussion of Jupiter in the "fifth house," these divine conditions apparently do not refer to a chart with Jupiter in that house. That is, the remarks are made completely independent of Jupiter's house. Augustus's horoscope had Jupiter in the tenth house (Midheaven). And if Firmicus was referring to a chart for April 17, 6 B.C., the time would have been sunset.

17. The mathematical techniques focused on predicting the planetary position along the ecliptic, namely, the longitude or the degree of the planet in a zodiacal sign. But none of the planets, except for the Sun, are on the ecliptic; the planetary orbits incline from the plane of the ecliptic. The latitude, degree north or south of the ecliptic, was virtually impossible to determine with rudimentary mathematics. Thus, the astrologers could not estimate how close planets actually passed and needed to observe the conjunction.

18. Sol and Mithras were said to be unconquerable because they were om-

nipotent: they controlled or overcame all cosmic forces. See Ulansey, *Mithraic Mysteries,* 95–103, 108–10.

19. Patrick Bruun, "The Disappearance of Sol from the Coins of Constantine," *Arctos* 2 (1958): 15–37.

20. Close inspection of these coins often reveals the *equinoctial cross* on the cosmic orb. This was the crisscross formed by the intersection of the zodiac circle and celestial equator. See Ulansey, *Mithraic Mysteries,* 131 n. 2; and Otto J. Brendel, *Symbolism of the Sphere* (Leiden: E. J. Brill, 1977), plate 17.

21. RPC-1 4286.

22. It is evident that RPC-1 4286 was replaced by RPC-1 4287 because the redesigned coin was used for the following two years, RPC-1 4290 and 4291. The new design was used for a three-year period (A.D. 55–58).

23. See the section in chapter 4 entitled "Regal Stars and Planets" for a discussion of the star and crescent symbol.

24. RPC-1 4804.

25. Suetonius *Nero* 6.

26. One can only wonder whether such astrological prognostications about Nero became self-fulfilling prophecies in a society that firmly believed in astrology.

27. Suetonius *Nero* 40; see another version in Dio Cassius *Roman History* 63.27.2. The passage in *Mathesis* 5.1.28 may refer specifically to Nero's horoscope.

28. Because the Sun was in the trine that it ruled and in the Ascendant, it was the ruler or lord of Nero's horoscope. It is likely that this is the reason that he associated himself with Helios throughout his life.

29. Tacitus *Histories* II.8.1. It is apparent that there were at least three Neronian impersonators.

30. Joseph Klausner, *The Messianic Idea in Israel: From Its Beginning to the Completion of the Mishnah,* trans. W. F. Stinespring (New York: Macmillan, 1955), 370. The *Sibylline Oracles* were issued from the second century B.C. to the fifth century A.D., and they are not related to the *Sibylline Books* that Romans consulted for prophecies.

31. Ibid., 371–81.

32. Rev. 13:18, 13:14.

33. Rev. 13:15. See Ernest Renan, *The Anti-Christ,* trans. W. G. Hutchinson (London: Hutchinson, 1899), 325–26. See also the discussion by Jack Lindsay, *Origins of Astrology* (New York: Barnes & Noble, 1971), 281–82.

34. Another explanation of the two-year period is that the third ruling planet of the trine of Aries, namely, Saturn, was present in Aries for about that time. Once Saturn left Aries there was no vestige of the rulers of the trine of Aries that had formed the Star of Bethlehem. However, the presence of Saturn in Aries did not by itself constitute a regal portent.

35. There were strong pagan religious and philosophical elements in Greek astrology. The moral principles were largely Stoic and Neoplatonic. Much of the pagan influence disappeared over the centuries. See Barton, *Ancient Astrology,* 30–31; Cramer, *Astrology in Roman Law,* 68; and Firmicus *Mathesis* 1.4, 1.7.

36. Will Durant, *The Story of Civilization,* vol. 3, *Caesar and Christ* (New York: Simon & Schuster, 1971), 559.

37. Acts 11:19–26.

38. Acts 7:2–56.

39. As for the significance of the shepherds in the Lucan account, Brown (*Birth of the Messiah,* 401–2, 420–24, 672–75) explains a number of theories. For instance, some researchers suggest that this reference points to a birth between March and November, when shepherds would be out in the fields around Bethlehem. The use of the shepherds could also be a midrash convention springing from Mic. 4:8 about the Tower of the Flock. Another theory is that the shepherds are an allusion to David, born in Bethlehem, who was a shepherd when he was anointed king. Perhaps most important, the shepherds play a role parallel to that of the Magi in the Matthean account in revealing the birth of Jesus. Ibid., 411–12, 444–45, 672. That use of the shepherds supports the theory that Luke was trying to tell the story without references to pagan astrology; thus, shepherds surrounded by the "heavenly host" replaced the Magi watching "his star."

40. See "Theories about the Star of Wonder" in chapter 2 for information about the "heavenly host," a term biblical scholars believe to have indicated the stars, particularly planets. This idea is supported by findings about Hellenistic Jews. According to Ness ("Astrology and Judaism"), there is strong evidence from around the fourth century A.D. that Jews began to accept astrology and adapted it to fit Judaic principles. While traditional Jews avoided astrology, others living in Hellenistic societies believed that the planets were doing the work of God under his control and guidance. Zodiacs began to appear in some synagogues, and prayers were offered to the planets, who were identified with angels, superhuman servants and messengers of God. The host of heaven were seen as obedient to the will of God and not as Greek planet-gods. These findings about Hellenistic Jews support the theory that the heavenly host in the account in Luke were, indeed, an allusion to a great celestial portent. By referring to the heavenly host, the account avoids acknowledging astrology, a practice frowned upon by traditional Jews.

41. Clement of Alexandria *Stromata* 1.21.145. Whether the Sun was exalted in Aries on April 20–21, 6 B.C., depends on how the Magi oriented the zodiac with respect to the vernal equinox (see Appendix A). If they used ♈ 5° = v.e., the Sun was in Taurus. But ♈ 3° = v.e. places it in Aries along with the other trine rulers, Jupiter and Saturn, which is, as we have seen, a powerful condition. The Moon, however, did not contribute to regal conditions on April 20–21, which produces a horoscope with less regal significance than April 17, 6 B.C. Another date from Clement, May 20, was relatively inauspicious for 6 B.C.

APPENDICES

1. For the problem in defining the vernal equinox, see Neugebauer, *Mathematical Astronomy,* 593–600, and especially O. Neugebauer and H. B. Van Hoesen, *Greek Horoscopes,* 179–83. See also Ptolemy *Tetrabiblos* 1.22. Ptolemy tells us that the power of the signs is derived from the equinoxes; thus, the signs should be referenced to these points, not the physical constellations.

2. Technically speaking, we can define a conjunction as a point, which has no angular breadth.

3. Ptolemy *Tetrabiblos* 2.8; 3.13.

4. Dorotheus *Pentateuch* 1.5; Hephaistio of Thebes *Apotelesmatics* 1.12.

5. Modern astrologers use a "Placidus" system, which maps the houses and cardinal points according their proper trigonometric positions.

6. T. Rice Holmes, "The Birthday of Augustus and the Julian Calendar," *Clas-*

sical Quarterly 6 (1912): 73–81. See also A. E. Houseman, "Manilivs, Avgvstvs, Tiberivs, Capricornvs, and Libra," *Classical Quarterly* 7 (1913): 109–14.

7. A problem in casting Augustus's chart lies in defining the position of the zodiacal signs relative to the vernal equinox. If we use the "standard" system that Claudius Ptolemy recommended, in which the vernal equinox marks the beginning of Aries (♈ 0° = v.e.), we find that the Moon falls in the twenty-ninth degree of Capricorn ♑ for September 23, 63 B.C. This position puts the Moon in its nocturnal trine and produces a powerful horoscope. We know from Dorotheus of Sidon that it was important for the dominant luminary (Sun during day, Moon at night) to reside in its own trine. Augustus was born "just before sunrise," which produces a nocturnal horoscope; therefore, the Moon must reside in Trine II (Taurus ♉, Virgo ♍, Capricorn ♑) if it is to rule Augustus's chart. But if we place the vernal equinox within Aries, that is, if we apply the correction ♈ 5° = v.e., the Moon moves into the fourth degree of Aquarius ♒. The chart loses this regal portent if the Moon is in Aquarius, where it would be for September 23 and ♈ 5° = v.e. We will use September 22 and ♈ 5° = v.e., although virtually everything we discuss is valid for a chart for September 23 and ♈ 0° = v.e. See the excellent article by Barton, "Augustus and Capricorn."

8. The software used for the celestial computations is W. C. Annala, LodeStar Plus (Zephyr Services: Pittsburgh, Penn., 1990). The results obtained with LodeStar Plus were checked with Jeffrey Sax's companion software to Jean Meeus, Astronomical Algorithms (Willmann-Bell: Richmond, Va., 1991).

9. Rumors in A.D. 11 of Augustus's imminent death forced him to publish his horoscope, which was probably accompanied by an "official" interpretation from Thrasyllus, who promised a longer life. Augustus died in A.D. 14.

10. Although the Sun is in the Ascendant, it is also in Libra, its fall, or depression. Perhaps having the Sun close to the border of Virgo (if we assume ♈ 0° = v.e., it was actually in Virgo) mitigated this adverse effect.

11. The presence and significance of the Moon in Capricorn for Augustus is supported by a remark by C. Fonteius Capito, close ally of Mark Antony and military escort to Cleopatra. Prior to the battle of Actium—which Antony and Cleopatra would lose, making Augustus the ruler of the Mediterranean world—Capito referred to a "tyrant" under "the Moon in Capricorn," which is most likely Augustus. See Cramer, *Astrology in Roman Law*, 67–68. This reference to the Moon agrees with the idea advocated by other researchers that Capricorn is a moon sign. The Moon resides in its own trine in Capricorn. Thus, referring to Capricorn as a moon sign greatly understates its significance.

12. The ruling Moon is followed by Mars in Taurus, which is in the same trine—another example of attendance.

13. Neugebauer and Van Hoesen, *Greek Horoscopes*, 90–91; and Cramer, *Astrology in Roman Law*, 165–70.

14. Hephaestion's birth chart is in Neugebauer and Van Hoesen, *Greek Horoscopes*, 131, chart L380. See also Hephaistio of Thebes *Apotelesmatics* (trans. Schmidt, ed. Hand) 1.

15. This quote is a combined interpretation of the translations given by Neugebauer and Van Hoesen, *Greek Horoscopes*, 90–91; and Cramer, *Astrology in Roman Law*, 164–70. The references to the moira, which only prolong the discussion of Hadrian's chart and are irrelevant to this discussion, have been removed. Antig-

onus may have been wrong that the Moon was in the first degree of Aquarius. It was in Capricorn, as it was in Augustus's horoscope. Nevertheless, Antigonus's interpretation of aspects, correct or not, is important in defining auspicious regal conditions.

16. Cramer, *Astrology in Roman Law,* 169.

17. See Tamsyn S. Barton, *Power and Knowledge: Astrology, Physiognomics, and Medicine under the Roman Empire* (Ann Arbor: University of Michigan Press, 1994), 76 passim.

Glossaries

Almanac: A compilation (usually yearly) of celestial, meteorological, agricultural, nautical, and civil information (see also *ephemeris*).

Angles: See *cardine*.

Anti-Midheaven: The lowest sign and the fourth mundane house, the one opposite the Midheaven. It is also called the *Imum Caelum* (or *Imum Caeli*).

Ascendant: (1) The point of the zodiac rising on the eastern horizon; also called the *Horoscopus*. (2) The first cardine sector in the system of mundane houses; extends from the Horoscopus to thirty degrees below the horizon. Generally, this is the rising sign.

Aspect: A geometric relationship between two or more celestial bodies in the zodiacal circle; includes conjunction, trine, sextile, quartile, and opposition.

Attendant: A planet that rises or sets after the Sun or the Moon. When Jupiter or Saturn rises before the Sun, or when Venus or Mars sets after the Moon, it is said to serve as an attendant, or "spear-bearer." Attendance has strong regal significance.

Benefic: A celestial body noted for its good influences (also called beneficent). Jupiter, Venus, and the Moon are benefics.

Bright star: A fixed star that held special powers similar to those of a planet. Four were believed to have powerful regal powers Aldebaran, Regulus, Antares, and Fomalhaut.

Cardinal points: The highest, lowest, rising, and setting signs (see *cardine*).

Cardinal signs: The zodiacal signs that include the equinoxes and solstices: Aries (vernal equinox), Cancer (summer solstice), Libra (autumnal equinox), and Capricorn (winter solstice).

Cardine: One of the four sectors, or zones, that cover cardinal points or axes for the surrounding sky. They include the Ascendant, Midheaven, Descendant, and anti-Midheaven. The cardines, also known as the angles, coincide with the first, fourth, seventh, and tenth mundane houses, respectively.

Conjunction: A configuration in which two or more celestial bodies lie close to each other, especially in the same sign.

Degree: An angular unit measuring 1/360th of a full circle. Each zodiacal sign covers 30 degrees. The twelve signs make up 360 degrees. The disks of the Sun and the Moon are each ½ degree across.

Depression: The position along the zodiac located opposite the exaltation. A planet or luminary in its depression was held to be weakest in influence (see *fall*).

Descendant: The cardinal point (*Dusis*) on the western horizon, or the cardine that extends upwards thirty degrees from the western horizon along the ecliptic. It is the seventh mundane house.

Domicile: A sign, or "house," assigned to a planet or luminary. A planet in its own house was assigned special influences. The Sun and Moon held one house, while the planets each held a pair. These houses are not related to mundane houses (places).

Ecliptic: The path of the Sun around the sky. When the Moon is on the ecliptic and in line with the Sun, eclipses will occur. The ecliptic is actually the earth's orbit projected against the background sky.

Epagomenal days: The five extra festival days added to the twelve thirty-day months needed to fill out the year in the old Egyptian calendar. A sixth was added during leap years in the new Egyptian calendar.

Ephemeris (ephemerides): Tabulations of calculated celestial and planetary positions for different times. These data are often included in an almanac, which has other civic information related to the calendar.

Equinoctial cross: The cross formed by the intersection of the ecliptic and equator. The cross was often inscribed on celestial orbs denoting the power of a world ruler (see *kosmokrator*).

Era: The system or convention used to enumerate the years. The Christian Era is now the most widely accepted convention.

Exaltation: Locations on the zodiac believed to hold omnipotent power for the planets and luminaries. Some astrologers specified a certain degree, but more commonly the entire sign was used.

Fall: The position along the zodiac opposite the exaltation. A planet or luminary in its fall loses its potential to influence the astrological conditions (see *depression*).

First angles: The two most important cardinal points or houses: the Ascendant and the Midheaven. Also called primary cardines or primary angles.

Heliacal rising: When a planet moves away from the Sun to become visible (astronomical definition). When a planet passes beyond the Sun's "arc of combustion" to exhibit enhanced powers (astrological definition).

Horoscopus: The place in the zodiac that crosses the eastern horizon. The Horoscopus defines the rising point and orientation of the zodiac for use in a horoscope, or horoscopic chart. It is also called the Ascendant, but the Ascendant was actually the rising sector (cardine), whereas the Horoscopus was a specific degree of the zodiac on the eastern horizon.

House: (1) A domicile. (2) One of the twelve sectors in the sky (mundane houses) surrounding the newborn child; controlled specific future events in the child's life.

Imum Caelum (Imum Caeli, IC, IMC): The anti-Midheaven, or subheaven. This is the lowest cardine and sign beneath the earth. It is the fourth mundane house.

Judicial astrology: An early form of astrology that focused on the fate of a king, government, or country rather than a specific person as in natal horoscopic astrology.

Latitude: In Greek astrology and astronomy, the angular distance (in degrees) of a star or planet from the ecliptic.

Longitude: In Greek astrology and astronomy, the angular distance (in degrees eastward along the ecliptic) of a star or planet from the vernal equinox. For a star or planet not exactly on the ecliptic, the point at which the star came closest to the ecliptic was used to determine the longitude.

Lord: The dominant celestial body that rules or controls a horoscope. Sometimes the title denotes a powerful ruling body in a significant aspect.

Lot: Position along the zodiac determined from the separations of celestial bodies and measured from the Ascendant (also called sorts). These predicted certain conditions for a person's life.

Luminaries: The Sun and the Moon.

Malefic: A celestial body noted for its bad influences (also called maleficent). Mars and Saturn are the malefics and were notorious for producing bad portents. When in the proper aspects with beneficent planets, they were believed to produce good influences.

Matutine: Literally, morning. Matutine rising means rising before the Sun.

Medium Caelum (Medium Caeli): The Midheaven, the tenth mundane house, and the highest sign. An extremely powerful, good cardine (see *Midheaven*).

Midheaven: The Medium Caelum; the tenth house; the highest cardine, the point of the zodiac that reaches the farthest north as viewed from the place of birth. The highest point is called the upper culmination.

Mundane houses: The twelve sectors that divide the sky around the newborn child (also called places). The first house is the Ascendant; the fourth, the subheaven; the seventh, the Descendant; and the tenth, the Midheaven. All were assigned control over the future life of the child.

Natal horoscope (or horoscopic chart): The astrological chart that represents the sky for the moment and place of a birth. The ancient charts were usually squares subdivided into the mundane houses that centered around the Horoscopus.

Native: The person whose time and place of birth is used to construct the horoscopic chart. The subject of the astrological analysis.

Occultation: The obscuration of one celestial body by another, especially the Moon. The term also refers to a celestial body that is hidden because it has set.

Opposition: An astrological aspect in which two celestial bodies are six signs apart, or about 180°. Opposition has a mixture of good and bad influences.

Places: These sectors are said to control future events for the native according to which celestial body occupies the sector (see *mundane houses*).

Planet: According to the ancients, anything that moved in the sky along the zodiac. This Greek word means "wanderer." Modern astronomers do not apply this term to the Sun and the Moon but use it only for the celestial bodies that orbit the Sun.

Precession: The drift of the earth's axis of rotation over the centuries. Precession is caused largely by the force of the Moon's gravity, which makes the earth wobble like a toy top.

Primary cardines: The two mundane houses that cover the Ascendant and the Midheaven. These are the first and ninth mundane houses, which were considered the most important, especially in regal portents. Also called first angles or primary angles.

Quartile (square): An astrological aspect in which two planets are ninety degrees, or three signs, apart. Bad portents are produced especially when maleficent planets, Mars and Saturn, are in quartile with the Moon or Sun.

Retrograde motion: The motion of a planet against the background stars in the same sense as the sky moves. The ancients called this "going before." Retrograde motion is essentially an optical illusion due to the earth's motion, which affects how we see a distant planet's motion among the constellations.

Ruler: A celestial body said to have enhanced powers and control over the astrological configuration. A ruler is similar to a lord, although it is possible to have several co-rulers in an horoscope.

Sect: An attribute of celestial bodies. The diurnal, or day, sect includes the Sun, Saturn, and Jupiter. The nocturnal, or night, sect include the Moon, Mars, and Venus. Mercury can be placed in either. The powers of the sect were believed to be accentuated according to whether it was day or night.

Sextile: An astrological aspect in which the celestial bodies are separated by two signs, or sixty degrees. The weakest of the aspects, it is a mildly good condition.

Sidereal: Pertaining to the sky or stars. The sidereal period is the time for a complete revolution of a celestial body relative to the stars. This is the true orbital period of a planet.

Sign: Not a constellation in the strict sense but a thirty degree sector of the zodiac—one of twelve measured from the vernal equinox. The Greek word for sign is *zoidion,* "a picture."

Solstice: The point were the Sun appears to stop and reverse its northward or southward direction. The northern limit marks the summer solstice, and the southern is the winter solstice.

Spear-bearers: Planets that served as protective attendants of the all-important luminaries, namely, the Sun and the Moon. Spear-bearers, which were believed to "hurl rays," of the same sect could produce a regal chart by guaranteeing the protection of the luminaries (see *attendant*).

Station: The place where a planet appears to stand still against the background stars before it reverses its direction (see *retrograde motion*).

Superior: Above, as in "above the horizon."

Terms: Sections within the signs assigned to each planet. The sectors in the prevalent Egyptian system of terms ranged from four to eight degrees in extent. A planet in its own term held special, enhanced powers and was almost as powerful one in its domicile.

Trine: An aspect in which two bodies are four signs, or 120°, apart. The trine was the most important aspect in Greek astrology.

Tropical: Anything that uses the equinoxes and solstices to describe or reference itself. A tropical year is the time required for the Sun to travel between successive vernal equinoxes.

Vespertine: Literally, evening. Venus is vespertine setting when it is seen in evening sky after sunset.

Zodiac: The circular band of constellations, or specifically signs, that covers the ecliptic.

HISTORICAL AND NUMISMATIC TERMS

Denarius (denarii): A standard silver coin issued by imperial mints. A denarius weighed about 3.8 grams and was close to 19 millimeters in diameter. Nero reduced the silver content from 98 to 93 percent.

Ekpyrosis: Conversion into fire; the Stoic belief that at the end of an age the world would be consumed by fire. In astrology a planet is consumed in the arc of combustion about the Sun. In either of these cases, *ekpyrosis* was followed by a rebirth, called *palingenesis.*

Equestrian: Ruled by a Roman knight (*eques*). Prefects of equestrian provinces were appointed by the emperor.

Ethnarch: A ruler with less stature than a king: a governor.

Ethnic: A phrase on a coin that designated the people of the issuing municipality.

Fortuna: The Roman goddess of fate; known as Tyche to the Greeks. Her symbols are the cornucopia and rudder.

Governor: Any ruling magistrate, but usually refers to a magistrate of a province. Under Augustus there were two types of provinces, imperial and public. Legates and proconsuls made up the governors of imperial provinces and served until they were recalled by the emperor. Public provinces were governed for one year by praetors and consuls. Egypt was uniquely governed by an equestrian prefect whom the emperor appointed and controlled.

Hasmoneans: The royal family of the ancient Jews.

Kosmokrator: A divine appellation for a world ruler such as Sol or Mithras; symbolized by the cosmic orb.

Legate: A high-ranking delegate or representative of the emperor. This title was given to very esteemed men put in charge of imperial provinces (see *governor*).

Midrash: Ancient Jewish method used to explain a biblical account. This can be either a literary style or the technique of using a myth to convey an underlying truth.

Obverse: The primary or most important side of a coin usually depicting a head of a deity or ruler; the origin of the term "heads."

Palingenesis: Rebirth; the Stoic belief that after a death through fire a new age was spawned. This idea is expressed in Greek astrology when a planet, emerging from the arc of combustion about the Sun, is reborn "in the east" with strong powers.

Prefect: A commander, magistrate, or governor of a minor province. In the case of Judea the prefects were subordinate to the governor of Syria. The prefect of Egypt reported directly to the emperor (see *governor*).

Procurator: Essentially the same as a prefect. The title was given to the Roman equestrian (knight) official who governed Judea after the death of Agrippa (see *governor*).

Reverse: The less important side of a coin commonly referred to as the "tails" side (see *obverse*).

RIC: The Roman Imperial Coinage (see bibliography).

RPC: Roman Provincial Coinage (see bibliography).

Tetrarch: Literally the governor of a quarter; that is, the governor of a small land.

Tyche: The goddess of fate for the Greeks (see *Fortuna*).

SIGNIFICANT HISTORICAL PERSONS

Agrippa, Marcus: Octavian's right-hand man. He led the naval victory at Actium against Mark Antony and Cleopatra in 31 B.C.; and he died in 12 B.C.

Alexander the Great: Macedonian king who conquered the whole Near East and spread Hellenistic culture. He died in Babylonia in 323 B.C.

Antigonus, Mattathias: Son of Aristobulus II. He held the Judean throne from 40 to 37 B.C., when Roman troops supporting Herod the Great pushed the Parthian forces from Judea. He was executed in 37 B.C.

Antipas: Son of Herod the Great. He became tetrarch of Galilee when his father died in 4 B.C. Caligula removed him in A.D. 39.

Antipater: Father of Herod the Great. The Romans trusted him and made him a prefect to watch over Hyrcanus. He was poisoned in 43 B.C.

Antony, Mark: Great Roman military leader and politician whose life produced many intriguing alliances. His alliance with Cleopatra against Octavian would prove to be his last and boldest adventure. He committed suicide in 30 B.C. before Octavian's troops reached Egypt, where he was with his wife, Cleopatra.

Archelaus: Became ethnarch of Judea when his father, Herod the Great, died in 4 B.C. He was deposed by Augustus in A.D. 6.

Augustus Caesar: The first Roman emperor. Born Octavian, he was given the name Augustus by the Senate in 27 B.C. He was Julius Caesar's adopted son. He died in A.D. 14.

Aurelian: Roman emperor (r. A.D. 270–75). He instated the Sun-worshipping cult of Sol Invictus (Unconquerable Sun) and established December 25 as the birthday of Sol.

Balbillus: Prefect of Egypt under Nero, but most famous as Nero's astrologer. He retired to Ephesus before the end of Nero's reign.

Brutus: Major conspirator in the assassination of Julius Caesar. Brutus died fighting Octavian and Mark Antony at Philippi in 42 B.C.

Cleopatra VII: Queen of Egypt who conspired with Mark Antony to seize control of the Mediterranean world. She and Mark Antony committed suicide in 30 B.C. after they lost the battle of Actium to Octavian.

Constantine the Great: First Christian Roman emperor (r. A.D. 306–37).

Dionysius Exiguus: Christian monk responsible for establishing the Christian Era system for enumerating the years in the calendar.

Dorotheus of Sidon: Author of the astrological treatise *Pentateuch,* presumably written in the late first century A.D.

Firmicus Maternus, Julius: Author of *Mathesis,* which was written in ca. A.D. 346. *Mathesis* is a major Latin compilation covering innumerable astrological conditions extracted from many documents.

Hephaestion of Thebes (also called Hephaistion or Hephaistio): An important astrologer of ca. A.D. 415. The first volume of his *Apotelesmatics* draws from Ptolemy and Dorotheus.

Herod the Great: King of Judea from 37 to 4 B.C. With Roman support, he ousted Mattathias Antigonus and the Parthians from Judea and its neighboring lands. Although he was ruthless, he instituted major civic projects in Judea and throughout his realm.

Herodotus: Greek father of history (ca. 480–ca. 425 B.C.). He traveled and reported on the history of Greeks and barbarians.

Jesus: The Jewish religious leader whose life and teachings are the basis of Christianity. Also called Christ, which means Anointed One (Messiah).

Josephus, Flavius: A commander in the first Jewish revolt. He surrendered to Vespasian and prophesied that the Roman would become emperor. His prophecy saved him, and he went on to be an important chronicler of Jewish history. He died in ca. A.D. 100.

Manilius, Marcus: Roman poet who wrote the *Astronomica* on astrology in ca. A.D. 15.

Nero: Succeeded Claudius as Roman emperor in A.D. 54. His reign started out well but led to tyranny, which produced a major revolt in Judea. He was deposed by the Senate, and his suicide in A.D. 68 led Rome into a civil war.

Pompey (the Great): A great Roman military leader who did much to extend Rome's control over the Near East. He is largely responsible for bringing Judea under Roman domination in 63 B.C. Beaten by Caesar in a great civil war, he lost his life seeking refuge in Egypt in 48 B.C.

Ptolemy III: Hellenistic king of Egypt (r. 247–222 B.C.) who instated the use of the leap-year system to maintain agreement with the seasons.

Ptolemy, Claudius: Great scholar of Alexandria, Egypt (ca. A.D. 110–70). His astrological treatise, the *Tetrabiblos,* was a standard reference in classical times. His *Almagest* stands as one of the great mathematical astronomy works.

Quadratus: Legate of Syria (A.D. 51–60).

Quirinius: Augustus's legate of Syria (A.D. 6 to after 7). He oversaw the annexation of Judea into the Roman province of Syria. The King James Version of the Bible refers to him as Cyrenius.

Silanus: Legate of Syria (A.D. 11–17). His name appears on some of the first coins of Antioch depicting Aries, the Ram.

Suetonius: Biographer of the Caesars (*Lives of the Twelve Caesars*). He was sacked in A.D. 121–22 allegedly for indiscretions with Emperor Hadrian's wife and perhaps for publicizing unflattering details about Caesar and Augustus.

Tacitus: Historian and orator. He was born in A.D. 56–57 and died early in Hadrian's reign (after A.D. 115). His *Histories* and *Annals* rank as the most important primary sources of early Roman history.

Thrasyllus: Astrologer and friend of Tiberius. He also served in Augustus's court, which initiated a line of astrological advisors in imperial courts. He died in A.D. 36, a year before Tiberius.

Tiberius: Roman emperor (r. A.D. 14–37). His avid interest in astrology did much to advance its practice in Rome.

Valens, Vettius: An important astrologer of the second century A.D. from Antioch. His *Anthology* is an extensive and significant astrological treatise and includes many examples of horoscopes.

Bibliography

Allison, D. C. "What Was the Star That Guided the Magi?" *Bible Review* 9, no. 6 (1993): 20–24, 63.

Anonymous of 379. *The Treatise on the Bright Fixed Stars.* Project Hindsight: Greek Track, vol. 2a. Trans. Robert Schmidt, ed. Robert Hand. Berkeley Springs, W.Va.: Golden Hind Press, 1993.

Baigent, Michael. *From the Omens of Babylon: Astrology and Ancient Mesopotamia.* London: Arkana, 1994.

Barrett, A. A. "Observations of Comets in Greek and Roman Sources before A.D. 410." *Journal Royal Astronomical Society Canada* 72, no. 2 (1978): 81–106.

Barton, Tamsyn. *Ancient Astrology.* New York: Routledge, 1994.

———. "Augustus and Capricorn: Astrological Polyvalency and Imperial Rhetoric." *Journal of Roman Studies* 85 (1995): 33–51.

———. *Power and Knowledge: Astrology, Physiognomics, and Medicine under the Roman Empire.* Ann Arbor: University of Michigan Press, 1994.

Bjorkman, Judith Kingston. "Meteors and Meteorites in the Ancient Near East." *Meteoritics* 8, no. 2 (1973): 91–132.

Boll, Franz. *Studien über Claudius Ptolemäus.* Leipzig: B. G. Teubner, 1894.

Bouché-Leclerq, A. *L'Astrologie Grecque.* 1899. Reprint, Bruxelles: Culture et Civilisation, 1963.

Brendel, Otto J. *Symbolism of the Sphere.* Leiden: E. J. Brill, 1977.

Brown, Raymond E. *The Birth of the Messiah: A Commentary on the Infancy Narratives in the Gospels of Matthew and Luke.* New York: Doubleday, 1993.

Brown, Robert K., and Philip W. Comfort. *The New Greek-English Interlinear New Testament.* Ed. J. D. Douglas. Wheaton, Ill.: Tyndale House, 1990.

Bruun, Patrick. "The Disappearance of Sol from the Coins of Constantine." *Arctos* 2 (1958): 15–37.

Budge, E. A. Wallis. *The Rosetta Stone.* New York: Dover Publications, 1989.

Bulmer-Thomas, Ivor. "The Star of Bethlehem—A New Explanation—Stationary Point of a Planet." *Quarterly Journal Royal Astronomical Society* 33 (1992): 363–74.

Bunson, Matthew. *A Dictionary of the Roman Empire.* New York: Oxford University Press, 1995.

Burke-Gaffnet, W. "Kepler and the Star of Bethlehem." *Journal Royal Astronomical Society Canada* (December 1937): 417–25.

Burnett, Andrew, Michel Amandry, and Pere Pau Ripollès. *Roman Provincial Coinage.* Vol. 1. London: British Museum, 1992.

Burnett, C., K. Yamamoto, and M. Yano. *Abu Ma'šar: The Abbreviation of the Introduction to Astrology.* Leiden: E. J. Brill, 1994.

Butcher, Kevin. *Roman Provincial Coins: An Introduction to the Greek Imperials.* London: Seaby, 1988.

Carey, M., A. D. Nock, J. D. Denniston, W. D. Ross, J. Wight Duff, and H. H. Scullard, eds., with the assistance of H. J. Rose, H. P. Harvey, and A. Sofer. *The Oxford Classical Dictionary.* London: Oxford University Press, 1950.

Caspar, Max. *Kepler.* London: Abelard-Schuman, 1959.

Chapman-Rietschi, P.A.L. "Venus as the Star of Bethlehem." *Quarterly Journal Royal Astronomical Society* 37 (1996): 843–44.

Clark, David H., John H. Parkinson, and Richard F. Stephenson. "An Astronomical Re-appraisal of the Star of Bethlehem—A Nova in 5 B.C." *Quarterly Journal Royal Astronomical Society* 18 (1977): 443–49.

Cramer, Frederick H. *Astrology in Roman Law and Politics.* Philadelphia: American Philosophical Society, 1954. Reprint, Chicago: Ares Publishers, 1996.

Cumont, Franz. *Astrology and Religion among the Greeks and Romans.* New York: Dover, 1960.

Dio Cassius. *Roman History.* Vols. 6–9. Trans. Ernest Cary. Loeb Classical Library. Cambridge: Harvard University Press, 1917–27.

Dorotheus. *Dorothevs Sidonivs: Carmen Astrologicalvm.* Trans. David Pingree. Leipzig: B. G. Teubner, 1976.

Durant, Will. *The Story of Civilization.* Vol. 3, *Caesar and Christ.* New York: Simon & Schuster, 1971.

Dworetsky, M. M., and S. J. Fossey. "Lunar Occultations of Jupiter and Saturn, and the Star of Bethlehem." *The Observatory* 118, no. 1142 (1998): 22–24.

Firmicus Maternus, Julius. *Ancient Astrology Theory and Practice: Matheseos Libri VIII by Firmicus Maternus.* Trans. Jean Rhys Bram. Park Ridge, N.J.: Noyes Press, 1975.

Freitag, Ruth S. *The Star of Bethlehem: A List of References.* Washington, D.C.: Library of Congress, 1979.

Gingerich, Owen. *The Great Copernicus Chase.* Cambridge, Mass.: Sky Publishing Corp. and Cambridge University Press, 1992.

———. "The Origin of the Zodiac." *Sky and Telescope* 67, no. 3 (1984): 218–20.

Grant, Michael. *Roman History from Coins: Some Uses of the Imperial Coinage to the Historian.* New York: Barnes & Noble, 1995.

———. *The World of Rome.* New York: Penguin Books, 1960.

Grant, Robert M. *Augustus to Constantine: The Emergence of Christianity in the Roman World.* New York: Barnes & Noble, 1996.

Halsberghe, Gaston H. *The Cult of Sol Invictus.* Leiden: E. J. Brill, 1972.

Hephaistio [Hephaestion] of Thebes. *Apotelesmatics.* Bk. 1. Project Hindsight: Greek Track, vol. 6. Trans. Robert Schmidt, ed. Robert Hand. Berkeley Springs, W.Va: Golden Hind Press, 1994.

Herodotus. *The History.* Trans. David Grene. Chicago: University of Chicago Press, 1987.

Ho Peng-Yoke. "Ancient and Mediaeval Observations of Comets and Novae in Chinese Sources." *Vistas in Astronomy* 5 (1962): 127–225.

Holmes, T. Rice. "The Birthday of Augustus and the Julian Calendar." *Classical Quarterly* 6 (1912): 73–81.

Houseman, A. E. "Manilivs, Avgvstvs, Tiberivs, Capricornvs, and Libra." *Classical Quarterly* 7 (1913): 109–14.

Hughes, David. *The Star of Bethlehem: An Astronomer's Confirmation*. New York: Walker and Company, 1979.

Humphreys, Colin J. "The Star of Bethlehem—A Comet in 5 B.C.—And the Date of the Birth of Christ." *Quarterly Journal Royal Astronomical Society* 32 (1991): 389–407.

Huxley, George. *The Interaction of Greek and Babylonian Astronomy*. Belfast: The Queen's University, 1967.

Jones, A.H.M. *Augustus*. New York: Norton, 1970.

Josephus, Flavius. *The Complete Works of Josephus: Flavius Josephus*. Trans. William Whiston. Grand Rapids, Mich.: Kregel Publications, 1993.

Justin. *Epitome of the Philippic History of Pompeius Trogus*. Trans. J. C. Yardley. Atlanta: Scholars Press, 1994.

Klausner, Joseph. *The Messianic Idea in Israel: From Its Beginning to the Completion of the Mishnah*. Trans. W. F. Stinespring. N.Y.: Macmillan, 1955.

Kohlenberger III, John R., ed. *The Precise Parallel New Testament*. New York: Oxford University Press, 1995.

Kronk, Gary. "A Large Comet in 135 B.C." *International Comet Quarterly* (January 1997): 3–7.

Lindsay, Jack. *Origins of Astrology*. New York: Barnes & Noble, 1971.

MacDonald, George. "The Pseudo-Autonomous Coinage of Antioch." *Numismatic Chronicle*, 4th ser., 4 (1904): 105–35.

Manilius, Marcus. *Astronomica*. Trans. G. P. Goold. Loeb Classical Library. Cambridge: Harvard University Press, 1992.

Martin, Ernest L. *The Birth of Christ Recalculated*. Pasadena, Calif.: Foundation for Biblical Research, 1980.

Martin, Luther M. *Hellenistic Religions: An Introduction*. Oxford: Oxford University Press, 1987.

Miles, Clement A. *Christmas Customs and Traditions: Their History and Significance*. New York: Dover, 1976.

Millar, Fergus. *The Roman Near East: 31 B.C.–A.D. 337*. Cambridge: Harvard University Press, 1993.

Mills, W. E., ed. *Mercer Dictionary of the Bible* (Macon, Ga.: Mercer University Press, 1990.

Molnar, Michael R. "Astrological Omens Commemorated on Roman Coins: Clues to Caesar's Fortune." *The Celator* 10, no. 3 (1996): 14–19.

———. "Astrological Omens Commemorated on Roman Coins: The Ides of March." *The Celator* 8, no. 11 (1994): 6–10.

———. "The Coins of Antioch." *Sky and Telescope* 83, no. 1 (1992): 37–39.

———. "The Magi's Star from the Perspective of Ancient Astrological Practices." *Quarterly Journal Royal Astronomical Society* 36 (1995): 109–26.

———. "Mithradates Used Comets on Coins as a Propaganda Device." *The Celator* 11, no. 6, (1997): 6–8

Mosley, John. *The Christmas Star*. Los Angeles: Griffith Observatory, 1987.

Ness, Lester J. "Astrology and Judaism in Late Antiquity." *The Ancient World* 26, no. 2 (1995): 126–33.

Neugebauer, Otto. *Astronomy and History: Selected Essays*. New York: Springer-Verlag, 1983.

———. *The Exact Sciences in Antiquity*. New York: Barnes & Noble, 1993.

———. *A History of Ancient Mathematical Astronomy*. New York: Springer-Verlag, 1975.

Neugebauer, O., and H. B. Van Hoesen. *Greek Horoscopes*. Philadelphia: The American Philosophical Society, 1987.

Neusner, Jacob. *A History of the Jews in Babylonia*. Vol. 1, *The Parthian Period*. Leiden: E. J. Brill, 1965.

O'Neil, W. M. *Time and the Calendars*. Parramatta, Australia: Sydney University Press and Macarthur Press, 1975.

Paffenroth, Kim. "The Star of Bethlehem Casts Light on Its Modern Interpreters." *Quarterly Journal Royal Astronomical Society* 34 (1993): 449–60.

Pannekoek, A. *A History of Astronomy*. New York: Dover, 1989.

Perowne, Stewart. *Herod*. New York: Dorset, 1991.

Peters, F. E. *The Harvest of Hellenism: A History of the Near East from Alexander the Great to the Triumph of Christianity*. New York: Barnes & Noble, 1996.

Pliny. *Natural History*. Bks. 1–2. Trans. H. Rackham. Loeb Classical Library. Cambridge: Harvard University Press, 1991.

Ptolemy, Claudius. *Tetrabiblos*. Trans. F. E. Robbins. Loeb Classical Library. Cambridge.: Harvard University Press, 1980.

Ptolemy's Almagest. Trans. G. J. Toomer. New York: Springer-Verlag, 1984.

Ramsey, John T., and A. Lewis Licht. *The Comet of 44 B.C. and Caesar's Funeral Games*. Atlanta: Scholars Press, 1997.

Renan, Ernest. *The Anti-Christ*. Trans. W. G. Hutchinson. London: Hutchinson, 1899.

Richardson, Peter. *Herod: King of the Jews and Friend of the Romans*. Columbia: University of South Carolina Press, 1996.

Rochberg-Halton, Francesca. "Elements of the Babylonian Contribution to Hellenistic Astrology." *Journal of the American Oriental Society* 108, no. 1 (1988): 51–62.

———. "New Evidence for the History of Astrology." *Journal of Near Eastern Studies* 43, no. 2 (1984): 115–40.

Sanders, Donald H., ed. *Nemrud Daği: The Hierothesion of Antiochus I of Commagene*. Winona Lake, Ind.: Eisenbrauns, 1996.

Schaefer, Bradley E. "Dating the Crucifixion." *Sky and Telescope* 77, no. 4 (1989): 374.

———. "Heavenly Signs." *New Scientist* (December 21–28, 1991): 48–51.

———. "Origin of the Star and Crescent." *Sky and Telescope* 83, no. 2 (1992): 129.

Shore, Fred B. *Parthian Coins and History: Ten Dragons against Rome*. Quarryville, Penn.: Classical Numismatic Group, 1993.

Silberman, Neil Asher. "Searching for Jesus: The Politics of First-Century Judea." *Archaeology* 47, no. 6 (1994): 31–40.

Sinnott, Roger W. "Thoughts on the Star of Bethlehem." *Sky and Telescope* 36, no. 6 (1968): 384–86.

Staal, Julius D. W. *The New Patterns in the Sky*. Blacksburg, Va.: McDonald and Woodward, 1988.

Suetonius. *Lives of the Caesars*. Trans. J. C. Rolfe. Loeb Classical Library. Cambridge: Harvard University Press, 1970.

Sutherland, C.H.V. *The Roman Imperial Coinage* Vol. 1. London: Spink, 1984.

Sutherland, C.H.V., D. Litt, and R.A.G. Carson. *The Roman Imperial Coinage.* Vol. 7. London: Spink, 1966.

Syme, Ronald. *The Roman Revolution.* London: Oxford University Press, 1963.

Tacitus. *Annals.* 3 Vols. Trans. J. Jackson. Loeb Classical Library. Cambridge: Harvard University Press, 1934, 1937.

———. *Histories.* 2 Vols. Trans. C. H. Moore. Loeb Classical Library. Cambridge: Harvard University Press, 1925, 1931.

Tester, Jim. *A History of Western Astrology.* New York: Ballantine, 1987.

Thomasson, Bengt E. *Laterculi Praesidum.* Arlöv: University of Gothenburg, 1984.

Thurston, Hugh. *Early Astronomy.* New York: Springer-Verlag, 1994.

Ulansey, David. *The Origins of the Mithraic Mysteries: Cosmology and Salvation in the Ancient World.* New York: Oxford University Press, 1989.

Valens, Vettius. *The Anthology.* 4 Bks. Project Hindsight: Greek Track, vols. 4, 7–8, 11. Trans. Robert Schmidt, ed. Robert Hand. Berkeley Springs, W.Va.: Golden Hind Press, 1993, 1994, 1994, 1996.

———. *Vettius Valens d'Antioche—Anthologies.* Bk. 1. Trans. Joëlle-Frédérique Bara. Leiden: E. J. Brill, 1989.

Viviano, B. T. "The Movement of the Star, Matt 2:9 and Num 9:17." *Revue Biblique* 103 (1996): 58–64.

Index

Abarbanel, Rabbi Isaac, 27–30, 77
Abraham, 104, 157n. 15
Abu Ma'šar, 152n. 32
Adonis, 30
Agrippa, Marcus, 21 table 2.1, 40, 59
    table 3.2
Aldebaran, 79
Alexander the Great, 34, 37–38, 52, 54
Amurru, 37–38
angels, 15, 122. *See also* heavenly host
angles, *see* cardinal points
Antares, 79
Antichrist, 109, 110, 115–116
Antigonus of Nicaea, 81, 89, 137
anti-Messiah, *see* Antichrist
anti-Midheaven, 27, 72–75, 132, 134;
    Moon in, 136; recovering lost pos-
    sessions in, 113
Antioch, 152n. 33; as capital of Syria,
    49, 53; Christian Church in, 120–
    121; coinage of, 3, 5, 48–51, 59, 86,
    109–111, 114, 120–124; Herod's
    projects in, 10; as home of evange-
    lists, 5
Antiochus I (king of Commagene),
    77–78
Antipas, Herod, *see* Herod Antipas
Antony, Mark, 8, 51
Aquarius, 21 table 2.1, 44
arc of combustion, 88
Archelaus, 48–49, 53, 59, 61
Aries: astrological symbolism of, 5;
    bright star in, 79; on coinage, 3,
    48–52, 59, 86, 110, 114, 120–125;
    equinoctial point, 26, 28, 44; exal-
    tation of the Sun, 72, 80–81, 98,
    106–107; geographic control of,
    46–47, 96, 111, 152n. 32; heliacal

risings in, 88–89, 94; Jupiter in, 27,
    93, 96, 102, 125; Jupiter's station in,
    94–95, 104; lunar occultations in,
    86; Mars in, 46, 129; mythological
    lore about, 49; and Nero's horo-
    scope, 109, 113; positioning the
    vernal equinox in, 127, 147n. 17;
    recast as a Christian symbol, 31; as
    regal sign, 75, 96; rulers of trine of,
    98–99, 107; as sign of the Jews, 5,
    47–48, 67–69, 109, 116–117, 152n.
    32; site of the Magi's star, 5, 53, 71,
    84, 97–103, 116; stationing planets
    in, 93; trine of, 29, 69
Aristotle, 22
Ascendant, 65, 72, 97, 132; angular dis-
    tance from the Midheaven line,
    134; attendant planets in, 75; bright
    star in Aries in, 79; bright stars in,
    79; first mundane house, 134; indi-
    cating stolen possession, 113;
    Jupiter and the Moon in, 81; Jupiter
    in Hadrian's horoscope in, 138;
    Moon in, 81, 100; position and ex-
    tent as a mundane house, 134; pri-
    mary angle, 72; regal conditions,
    72; Sun in, 75, 81, 111, 136; Sun or
    Moon in, 74
aspects, 128–129. *See also* conjunction;
    opposition; quartile; trine
attendant planets, 73–76, 102, 107; on
    April 17, 6 B.C., 99; in Augustus's
    horoscope, 136; in Hadrian's horo-
    scope, 81, 137–138
Augustus Caesar (Octavian) (Roman
    emperor), 3, 21 table 2.1, 40, 58;
    annexes Judea, 49, 51, 59; birth of,
    134; censuses by, 61; as divine,

Augustus Caesar (*continued*)
109; horoscope of, 40, 105–106,
134–137, 160n. 7; supports Herod,
8; taxation under, 60; as unconquerable general, 108; use of Capricorn
as natal sign, 3, 48; veneration of
the Julian comet, 18
Aurelian (Roman emperor), 56

Babylonia, 8, 34–39, 42, 52; my-
thology from, 78–80
Balaam, 6–7, 16, 36, 116, 118
Balak (king of Moab), 6–7
Balbillus, Tiberius Claudius, 116; inter-
pretation of Nero's comet, 20
Barrett, A. A., 20
Ben Yohai, Rabbi Simeon, 149n. 7
Berenice (queen of Egypt), 54
Bethlehem: birth of Jesus in, 119; as
city of David, 14; family of Jesus in,
62; origin of the Star of, 1
Bible: astrology in, 60, 118; versions of
New Testament, 5, 116. *See also*
Luke; Mark; Matthew
bright stars, *see* regal bright stars
Bulmer-Thomas, Ivor, 89

calendar: Babylonian, 38; Gregorian,
55; Julian, 53–55, 134; new Egyptian, 54; old Egyptian, 53–54;
Philocalian, 56
Caesar, Julius, 40, 53, 134; adopts leap
year calendar, 54; assassination of,
8; comet of, 18
Caesarea, 10
Cancer: exaltation of Jupiter in, 106,
135–136; summer solstice in, 44
Capricorn: coinage depicting, 40, 48;
Moon in, 136; as natal sign of Augustus, 3; winter solstice in, 44
cardinal points, 46, 72–74, 97, 99, 113,
132–136; importance of for conjunctions, 26; primary or first, 81.
See also anti-Midheaven; Ascendant; Descendant; Midheaven
census in Gospel of Luke, 49, 58–62,
120–121, 152n. 36, 154nn. 61–62
Chaldeans, as astrologers, 25, 35–36, 130

Chinese astronomical records, 24
Christ, *see* Jesus; Messiah
Christian Jews, 5, 120
Christian symbolism, 28, 31
Christmas, 1, 25, 56–57, 125
Church, early Christian, 36, 55,
120–121
Clement of Alexandria, 36, 55, 122
Cleopatra VII (queen of Egypt), 8, 21
table 2.1, 51
Coele Syria, 46–47, 49, 52, 111
coinage: Aries on, 50–53, 109–111;
Capricorn on, 3, 40; Julian comet
on, 18; Nero on, 114; provincial,
48; Sol Invictus on, 108–109
comets: Aristotelian theory of origin of,
22; as explanation of Magi's star, 1,
17–21, 26, 85, 135; and Mithridates
the Great, 146n. 6; during Nero's
reign, 19; unusual veneration of
Julian comet, 18
Commagene, 77, 151n. 29
conjunction, 1, 25–29, 71, 129,
155n. 6; believed to produce comets, 22. *See also* triple conjunction
Constantine the Great, 108; vision at
Milvian Bridge, 105, 157n. 15
Coponius, 49
cosmic orb, 109. *See also* equinoctial
cross
Cyrenius, *see* Quirinius
Cyril of Alexandria, Saint, 36

Damascus, 47, 111, 114
Darius (king of Persia), 53
David (king of Israel), 6, 14, 119
Dead Sea Scrolls, 11, 35
December 25, 55–57
Decree of Canopus, 54
Descendant, 72, 74, 132, 134
Dionysius Exiguus, 55, 57
divine portent, 104
domicile, 129, 132. *See also* house
Dorotheus of Sidon, 74, 80, 113, 133, 136

east, the: Jupiter in, 103; Jupiter in Hadrian's horoscope, 89, 137; Jupiter
on April 17, 6 B.C., 89, 103; Magi's

star in, 1, 3, 85, 116; meaning of in Greek astrology, 87. *See also* heliacal rising
eclipse, 57, 58, 84
ecliptic, 42, 46, 79, 127
Egypt, 6, 8, 34, 44, 47, 52–55, 88
ekpyrosis [consumption by fire], 88, 148n. 26
epagomenal days, 53
equinoctial cross, 158n. 20
equinoctial points, 26
era, 53, 55, 57, 128
exaltation, 67–72; importance of, 129; Jupiter's, 106, 135; of planets for April 17, 6 B.C., 98, specific degree of, 154n. 3; Sun's, 72, 81, 95, 107; of Venus in Hadrian's horoscope, 137

fate, 32–34, 110, 133
Firmicus Maternus, Julius, 67; on aspects of divine births, 104, 106; on bright stars, 77, 79; Christian beliefs of, 104–105, 109, 157n. 15; on elements of April 17, 6 B.C., 108, 119; on exalted Sun in Midheaven, 72; on heliacal risings, 88; on loss of inheritance, 112–113; on lunar conjunctions and occultations, 82, 106; prognosis for type of triple conjunction in 6 B.C., 27; on regal births, 74, 75, 79, 80, 82, 135; writes the *Mathesis*, 108
Fomalhaut, 79

Gemini, 43
Golden Fleece, 50
governor: and legate of Syria, 49–52, 58–61, 110; portent for birth of, 73, 75, 102; of the whole world, 82
Greek astrology, 20, 31, 64–67, 80–83, 87, 110–111, 116–118, 129; ambiguity of statistical analysis of, 102; origin of, 39; portents for regal births in, 42, 64, 77, 97; as practiced by the Magi, 42; primary sources for, 47; star and crescent symbol in, 78
guest star, 24

Hadrian (Roman emperor): horoscope of, 65, 66, 81–83, 89, 137–138; similarity of horoscope to that of April 17, 6 B.C., 99–100
heavenly host, 121, 122, 125. *See also* angels
heliacal rising, 87–89, 92–94, 100; of Jupiter on April 17, 6 B.C., 89; of Sirius (Sothis), 53
Helle, 50
Hephaestion (Hephaistio) of Thebes, 58, 73, 133, 137
Herod (king of Judea): ambitious building program of, 10; audience with Magi, 1, 3, 11, 14, 16, 89; death of, 48, 55–58, 117, 118, 122; extent of kingdom, 47; fights the Parthians, 8; kills his sons, 10, 62; messianic fervor during reign of, 8, 11; similarity to King Balak, 7; and Slaughter of the Innocents, 62
Herod Antipas, 49, 154n. 62
Herodotus, 33
Hipparchus, 24, 127
horoscope: construction of, 39, 65; earliest example of, 38; for April 17, 6 B.C., 97, 101, 109; for April 4, A.D. 54, 102; Nemrud Daği, 77. *See also under* Augustus Caesar; Hadrian; Nero
Horoscopus, 40, 72, 134, 137
house: domicile, 68, 106, 129; mundane, 27, 132–134. *See also* cardinal points
Hyrcanus II (king of Judea), 8

Idumea, 10, 46
immortal portent, *see* divine portent
Imperator, 114
infancy narratives, 5. *See also* Luke; Matthew

James, Protoevangelium of, 145n. 1
Jerusalem: flight of Christian Jews from, 120; inhabitants of did not recognize Magi's star, 11; as Messiah's capital, 119; as Nero's new capital, 110, 115–116; retaking of

Jerusalem (*continued*)
  by Herod, 10; under the sign of
  Aries, 86
Jesus, 28; birth of, 1, 5–7, 14–15, 49,
  57–63, 127; Crucifixion of, 5, 55,
  120; Herod's plan to kill, 7; ministry
  of, 60
Jews: and faith in advent of Messiah,
  16; and hatred for Herod, 10; and
  interest in astrology, 159n. 40; and
  nonpractice of astrology, 16, 35, 39;
  Roman interactions with, 8–11; and
  support of the Parthians against
  Rome, 8
John the Baptist, 60
Josephus, Flavius, 57–58, 62
Judea, 60; annexation of in A.D. 6, 49,
  120; economic success under
  Herod, 10; as focus of the Magi's attention, 11, 14, 27; as home of the
  Messiah, 8; incorporated into Syria
  Palaestina, 47; messianic fervor in,
  10; represented by Aries, 46
judicial astrology, 37
Jupiter: in Augustus's horoscope, 137;
  in exaltation, 106, 135–136; in Hadrian's horoscope, 138; heliacal rising of, 88, 101; lunar occultations
  of, 106; as Magi's star, 86–89, 93,
  96, 103, 117; in Midheaven, 136;
  three conjunctions of with Saturn
  in 7 B.C., 29; in triple conjunction,
  22–23, 27; as regal planet, 76, 79–
  83, 100; as ruler of Aries trine, 69;
  as star of Zeus, 25; stations of,
  93–94

Kepler, Johannes, 26–30; beliefs about
  the Magi's star, 23; observations of
  the 1604 supernova, 22; observations of the triple conjunction of
  1604, 25
King of the Jews, 7, 11, 48
kosmokrator [world ruler], 74, 108–
  109, 153n. 46

legate, *see* governor
Leo: regal bright star Regulus in, 77–
  79; in trine of Aries, 69

Libra: as equinoctial point, 26; role of
  in Augustus's horoscope, 106
loss of possessions, portents of, 112
Lot of Fortune, 154n. 1
lots, system of, 154n. 2
Luke, Gospel of: chronological information about Jesus in, 59, 119; infancy narrative of, 5; reconstruction
  of the Nativity, 122; references to
  Quirinius in, 49, 58, 119, 121; when
  written, 120

Magi: account of in Matthew, 3; audience with Herod, 14; country of
  origin, 36, 149n. 7; description
  of star by, 16; gifts of, 149n. 6;
  meaning of story about, 6; nature
  of star of, 17; origin of name, 32;
  parallels of Matthew's story with
  story of Balaam, 7; practice of
  Greek astrology, 31, 33; report
  two astrological portents, 96; star
  of, 15; visit of, 11
magi: origin of, 33; and practice of astrology, 33, 36; as wise men, 42
Manilius, Marcus, 33, 47
Mark, Gospel of, 5
Mars: destructive nature of, 20, 76, 102;
  domicile in Aries, 46; in Nero's
  horoscope, 111; in triple conjunction, 22, 27
Mary, mother of Jesus, 31, 36
Matthew, Gospel of: astrological report
  in, 92–93, 96, 117; description of
  star in, 16; historicity of, 116, 122;
  infancy narrative of, 5, 14, 16; and
  midrash interpretation about the
  Magi, 7; as sole source for Magi's
  star, 1; when written, 120
Mercury: exalted in Augustus's horoscope, 136; fickle nature of, 20
Mesopotamia, 8, 36, 115
Messiah, 115; birth of, 6; Jesus as, 6,
  7, 120; in lineage of David, 14;
  prophecy of, 6; Roman fear of, 121
meteors, 37
Midheaven, 66; regal portents in, 72,
  80–81, 99–100, 134
midrash, 6–7, 119, 123

miracle theory, 1, 6, 16–17, 26; Kepler's belief in, 23–24
Mithras, 56, 108
Mithridates the Great (king of Pontus), 146n. 6
Moon: Babylonian studies of, 38; in Capricorn in Augustus's horoscope, 136; and conjunctions with bright stars, 77, 79; in Hadrian's horoscope, 137; influence of in triple conjunction, 27; Midheaven influences of, 27; and regal conjunctions with Jupiter, 81; and occultations with Jupiter, 82, 83
morning phase, *see* heliacal rising
morning rising, *see* heliacal rising
morning star, 87–88, 92, 99, 116, 122, 138; as messenger of God, 16
Moses: birth of, 28; and story of Balaam, 6
Muhammad, 78

Nabatea, 10
Nativity, *see* Jesus, birth of
Nemrud Daği, 77–78
Nephele, 50
Nero (Roman emperor), 19–20; horoscope of, 110–116
Neugebauer, Otto, 83, 128
new star, 21–24. *See also* nova; supernova
nova, 21, 127

occultation: on April 17, 6 B.C., 107; astrologers' interest in, 83; effect of lunar nodes on, 156n. 2; of Jupiter by the Moon, 86
Olympic Games, 10
Ophiuchus, 22
opposition, 100, 129, 137

Palestine, 46, 152n. 32
palingenesis [rebirth], 88, 148n. 26
Parthians, 8, 11, 36, 40; known for kings with long hair, 18; support for Jews, 149n. 7
Passover, 57–58
Paul the Apostle, 5, 120
Persians, 32–33, 149n. 7, 152n. 32

Philip (son of Herod), 49
Phoenicia, 47
Phrygian cap of the Magi, 36
Pisces, 44, 148n. 22; recast as a Christian symbol, 28; site of the 6 B.C. triple conjunction, 26
Piso, L. Calpurnius, 59
planetary conjunctions, 1; as explanation of Magi's star, 25–30
Pliny the Elder, 18, 24
precession of the equinoxes, 44, 56, 127–128
prefect of Judea, 49, 154n. 62. *See also* procurator
primary angles, 72. *See also* Ascendant; Midheaven
procurator, 115, 145n. 10
proskynesis [worship], 81, 83
Ptolemais, 114
Ptolemy III (king of Egypt), 54
Ptolemy, Claudius: analyzes heliacal risings, 88; compiles the *Tetrabiblos*, 44; describes imperial births, 74; does not relate Pisces to Judea, 28; as primary source on Greek astrology, 20; relates Aries to Judea, 46

Q, biblical source, 6
Quadratus, C. Ummidius Durmius, 110
quartile, 100, 102, 112, 129, 157n. 14
Quirinius, Publius Sulpicius: called Cyrenius in Bible, 58; contact with Tiberius and Thrasyllus, 152n. 37; directs annexation of Judea, 49; duration as legate, 59; possibly issues Aries coinage, 49, 51; taxation system implemented by, 60
Qumran, 11

rarity of events, 17, 26, 66, 71, 101–102
regal births: defined, in Greek astrology, 42, 64; interest of astrologers in, 74; interest of Magi in, 14. *See also* regal bright stars; regal principles
regal bright stars, 77, 79
regal principles, 67–69, 76–77, 84, 97, 101–102, 136

Regulus, 77–79
retrograde motion, 90–96, 104, 117.
    See also stood over; went before
Revelation of Saint John the Divine,
    115–116
Rosetta Stone, 54

Sagittarius: as part of the trine of Aries,
    29; role of on December 19, 6 B.C.,
    95; as site of Kepler's triple conjunc-
    tion, 22, 26
Samaria, 47–49, 60, 121, 154n. 62
Sanders, Donald, 78
Saturn: as co-ruler of Aries trine, 69;
    destructive nature of, 76, 112; pur-
    portedly related to the Jews, 29,
    148n. 25; three conjunctions of
    with Jupiter in 7 B.C., 29; triple con-
    junctions of, 22–23, 27
Saturnalia, 56
Saturninus, S. Sentius, 61
Scorpius: Hipparchus's nova in, 24; re-
    gal bright star Antares in, 79; and
    sign called Scorpio, 44
sect, 74, 112–113, 131
Seleucids, 37, 52
Seneca, Lucius Annaeus, 19, 20, 34
shepherds, in Gospel of Luke, 49, 121,
    125, 159n. 39
Sibylline Oracles, 35, 115
sidereal year, 54
signs, *see* zodiacal signs
Silanus, Q. Caecilius Metellus Creticus,
    51–52
Sirius (Sothis), 53–54
Slaughter of the Innocents, 62,
    117–118
Sol Invictus, 56–57, 108–109
Sosigenes, 53
Sothic Cycle, 54
spear-bearer, *see* attendant planets
star and crescent, 78, 110–111, 114
stations, 93. *See also* Jupiter, stations
    of; retrograde motion; stood over
statistical rarity of events, *see* rarity of
    events
Stephen, Saint, 120

Stoicism, 34, 56
stood over, 15, 87–89, 92–93, 96, 104,
    117
Suetonius, 18, 20, 40, 110–114
Sun: astrological effects of, 46; in Au-
    gustus's horoscope, 136; and com-
    bustion of planetary powers, 87;
    defines the ecliptic, 42; exalted in
    Aries, 69; in Hadrian's horoscope,
    137; in Nero's horoscope, 111; role
    in triple conjunction of, 27; as ruler
    of Aries trine, 69
supernova, 1, 21–22, 30, 85, 127, 135;
    as explanation of Magi's star, 21–
    25; 1604 appearance of, 23
Syria: as distinct from Coele Syria, 47;
    as home of the evangelists, 5; in-
    vaded by Parthians, 8; zodiacal
    sign of, 151n. 29
Syria Palaestina, 47

Tacitus, 19, 29, 148n. 25, 152n. 36,
    158n. 29
Taurus, 79, 93
Temple of Solomon, 10, 145n.10
terms, system of, 129
Tertullian, 33, 61
Tester, Jim, 104
Tetrabiblos, 21, 44–47, 74, 92–93, 112,
    124. *See also* Ptolemy, Claudius
Theogenes, 40
Thrasyllus, Tiberius Claudius, 151n.
    29, 152n. 37
Tiberius (Roman emperor), 60, 119
Trier, 109
trigon, fiery, 29
trine, 29, 69, 98, 138; rulers of, 102
triple conjunction, 26–30; in 6 B.C.,
    25–26; 1604 supernova appears in
    midst of, 22
Tyre, 10

unconquerable, 56, 106, 108

Valens, Vettius: analysis of triple con-
    junctions, 27; confirms Aries' con-
    trol of Herod's kingdom, 47;

prognosis for the type of triple conjunction in 6 B.C., 26–27

Venus: exalted in Hadrian's horoscope, 137–138; exalted on April 17, 6 B.C., 98; lunar conjunctions with, 83; role in Augustus's horoscope of, 137; as star of Aphrodite, 25

vernal equinox, 26, 44, 127, 130; location of, 128, 130, 147n. 17, 159n. 41, 160n. 7; precession of, 28, 127

Vespasian (Roman emperor), 18

Vienna (Gaul), 49

Virgo: recast as a Christian symbol, 31; role in Augustus's horoscope of, 136

went before, 15, 87–96, 104, 117

white dwarf star, 22

Wise Men, *see* Magi

world ruler, *see* kosmokrator

zodiac: astrological powers from, 44; depicted in some synagogues, 35; effect of precession on, 128; meaning of, 42; orientation in sky, 42; origin of, 38; role of comets in, 21

zodiacal signs: assignment of control over countries, 46; origin of, 38; revising mythology about, 31

Zoroastrians, 32–33

About the Author

Michael R. Molnar, PhD. is a retired astronomer. Currently, he is a luthier, making violins and writing about the physics underlying their construction.

CPSIA information can be obtained at www.ICGtesting.com
Printed in the USA
BVOW08s2357090913

330698BV00001B/3/P